李林燕　胡伏原　吕　凡　著

面向变化场景的连续人工智能

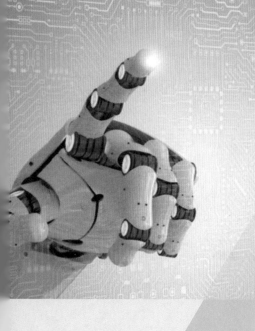

Continual
Artificial
Intelligence
towards
Changing
Environment

图书在版编目（CIP）数据

面向变化场景的连续人工智能 ＝ Continual Artificial Intelligence towards Changing Environment：英文 / 李林燕，胡伏原，吕凡著. -- 苏州：苏州大学出版社，2024.12. -- ISBN 978-7-5672-5015-4

Ⅰ．TP18

中国国家版本馆 CIP 数据核字第 20248Z5W57 号

Mianxiang Bianhua Changjing de Lianxu Rengong Zhineng

书　　名：	面向变化场景的连续人工智能
	Continual Artificial Intelligence towards Changing Environment
著　　者：	李林燕　胡伏原　吕　凡
策划编辑：	史创新　周建兰
责任编辑：	沈　琴　征　慧
责任校对：	倪锈霞　项向宏
装帧设计：	吴　钰
出版发行：	苏州大学出版社（Soochow University Press）
社　　址：	苏州市十梓街 1 号　邮编：215006
印　　刷：	镇江文苑制版印刷有限责任公司
邮购热线：	0512-67480030
销售热线：	0512-67481020
开　　本：	710 mm×1 000 mm　1/16　印张：12　字数：197 千
版　　次：	2024 年 12 月第 1 版
印　　次：	2024 年 12 月第 1 次印刷
书　　号：	ISBN 978-7-5672-5015-4
定　　价：	58.00 元

图书若有印装错误，本社负责调换
苏州大学出版社营销部　电话：0512-67481020
苏州大学出版社网址　http://www.sudapress.com
苏州大学出版社邮箱　sdcbs@suda.edu.cn

Preface

Continual learning allows a machine learning system to be adaptive in changing environment, which is an important research direction in artificial intelligence in recent years, and one of the core technologies for general artificial intelligence. However, the current literature on continual learning is scattered and incoherent, and readers interested in this field have to make a lot of effort to review many kinds of literature to get a general understanding of the research of this field. We have studied continual learning for years, and have had nearly 5 years of research experience so far. We are looking forward to integrating our recent 5 years' research work to facilitate readers' understanding of this field.

Based on our research work and related achievements in recent years, this book mainly summarizes the scenarios, difficulties, solutions of continual learning, and some applications of continual learning in multi-label classification problems, edge computation scenarios and few-shot classification. The study of continual learning will enable artificial intelligence to cope with changing data at a small cost in various scenarios. For example, in the medical field, the continual identification of variations of various diseases is one of the core problems of intelligent medical treatment. However, due to problems such as data privacy and rapid changes in viruses, the past artificial intelligence has been unable to support it for a long time. On the basis of this book, readers can quickly understand the development context of relevant fields and construct artificial intelligence applications in changing environment. In the process of introducing our research work, we interspersed some basic concepts and methods related to artificial intelligence. This is not only helpful for readers to gradually get familiar with the relevant professional knowledge in this field during the reading process but also helpful for readers to use it in their future work.

The main content of this book is the theory of continual learning and some key techniques and methods of its application. Our contributions include: (1) proposing a multi-domain multi-task rehearsal selection strategy and a two-level angular margin loss; (2) evaluating example influence in continual learning and proposing a better way to select rehearsal examples; (3) proposing an asymmetric gradient distance for parallel continual learning; (4) proposing an augmented graph neural network for continual multi-label classification; (5) proposing a federated continual learning algorithm based on accumulated Fisher information; (6) proposing a centroid-based rehearsal algorithm and constructing centroid distance loss; (7) applying continual learning to V2X perception application. This book aims to enable readers to have a more comprehensive understanding of the theory and applications of continual artificial intelligence.

Contents

Chapter 1 Introduction 1
  1.1 Static and Dynamic Artificial Intelligence 3
  1.2 Theory of Continual Learning 6
    1.2.1 Scenarios of Continual Learning 7
    1.2.2 Challenges of Continual Learning 9
    1.2.3 Approaches of Continual Learning 10
  1.3 Content of This Book 12

Chapter 2 Multi-Domain Multi-Task Rehearsal for Continual Learning 14
  2.1 Introduction 15
  2.2 Methodology 17
    2.2.1 Multi-Domain Multi-Task Rehearsal 17
    2.2.2 Two-Level Angular Margin Loss 20
    2.2.3 Episodic Distillation 22
    2.2.4 Total Algorithm 23
  2.3 Experiments 23
    2.3.1 Experimental Settings 23
    2.3.2 Comparison with the State-of-the-arts 24
    2.3.3 Domain Shift Observation 28
  2.4 Chapter Conclusion 30

Chapter 3 Exploring Example Influence in Continual Learning 31
  3.1 Introduction 31
  3.2 Methodology 34
    3.2.1 Preliminary: Rehearsal-based CL 34
    3.2.2 Example Influence on Stability and Plasticity 34

3.3　Meta Learning on Stability and Plasticity ……………………… 35
　　3.3.1　Influence Function for SP ……………………………………… 35
　　3.3.2　Simulating IF for SP …………………………………………… 36
3.4　Using Influence for Continual Learning ………………………… 38
　　3.4.1　Before Using: Influence for SP Pareto Optimality ………… 38
　　3.4.2　Model Update Using Example Influence …………………… 40
　　3.4.3　Rehearsal Selection Using Example Influence ……………… 40
3.5　Experiments ………………………………………………………… 41
　　3.5.1　Datasets and Implementation Details ……………………… 41
　　3.5.2　Main Comparison Results …………………………………… 42
　　3.5.3　Analysis of Dataset Influence on SP ………………………… 46
　　3.5.4　Analysis on SP Pareto Optimum …………………………… 47
　　3.5.5　Training Time ………………………………………………… 48
3.6　Chapter Conclusion ………………………………………………… 48

Chapter 4　Measuring Asymmetric Gradient Discrepancy in Parallel Continual Learning ……………………………………………… 50

4.1　Introduction ………………………………………………………… 50
4.2　Methodology ………………………………………………………… 53
　　4.2.1　Parallel Continual Learning ………………………………… 53
　　4.2.2　Measuring Asymmetric Gradient Discrepancy …………… 55
　　4.2.3　Maximum Discrepancy Optimization ……………………… 57
4.3　Experiments ………………………………………………………… 60
　　4.3.1　Dataset ………………………………………………………… 60
　　4.3.2　Experiment Details …………………………………………… 61
　　4.3.3　Main Results ………………………………………………… 62
　　4.3.4　Rehearsal Analysis in PCL …………………………………… 64
　　4.3.5　Comparison with Symmetric Metrics ……………………… 65
　　4.3.6　Ablation Study ………………………………………………… 66
　　4.3.7　Procedure Time ……………………………………………… 67
　　4.3.8　Tolerance Analysis in AGD ………………………………… 67
4.4　Chapter Conclusion ………………………………………………… 68

Chapter 5 Multi-Label Continual Learning Using Augmented Graph Convolutional Network ... 70

- 5.1 Introduction ... 71
- 5.2 Methodology ... 75
  - 5.2.1 Definition of MLCL ... 75
  - 5.2.2 MLCL Scenarios ... 76
  - 5.2.3 Overview of the Proposed Method ... 77
  - 5.2.4 Partial Label Encoder ... 79
  - 5.2.5 Augmented Correlation Matrix ... 81
  - 5.2.6 Objective Function ... 83
- 5.3 Experiments ... 86
  - 5.3.1 Datasets ... 86
  - 5.3.2 Evaluation Metrics ... 88
  - 5.3.3 Implementation Details ... 89
  - 5.3.4 Baseline Methods ... 89
  - 5.3.5 Main Results ... 90
  - 5.3.6 More MLCL Settings ... 93
  - 5.3.7 mAP Curves ... 94
  - 5.3.8 Ablation Studies ... 95
  - 5.3.9 Visualization of ACM ... 97
- 5.4 Chapter Conclusion ... 99

Chapter 6 Towards Long-Term Remembering for Federated Continual Learning ... 100

- 6.1 Introduction ... 100
  - 6.1.1 Federated Learning ... 102
  - 6.1.2 Federated Continual Learning ... 103
- 6.2 Methodology ... 103
  - 6.2.1 Problem Definition ... 103
  - 6.2.2 Multi-Node Collaborative Integration for Parameter Co-Importance ... 104

    6.2.3 Fisher Accumulating and Balancing for Reducing Forgetting 106

 6.3 Experiments 108

    6.3.1 Experiment Details 108

    6.3.2 Results 109

    6.3.3 Ablation Experiment 111

    6.3.4 Fisher Visualization 113

 6.4 Chapter Conclusion 114

Chapter 7 Centroid-based Rehearsal in Online Continual Learning 115

 7.1 Introduction 115

 7.2 Methodology 117

    7.2.1 Continual Domain Shift in OCL 117

    7.2.2 Centroid-based Rehearsal 118

    7.2.3 Distillation on Centroid Distance 121

    7.2.4 The Overall Algorithm 123

 7.3 Experiments 124

    7.3.1 Dataset and Experimental Details 124

    7.3.2 Experimental Results 125

 7.4 Chapter Conclusion 129

Chapter 8 Dynamic V2X Perception from Road-to-Vehicle Vision 130

 8.1 Introduction 130

 8.2 Methodology 136

    8.2.1 Overview 136

    8.2.2 Overcoming Intra-Scene Changes 137

    8.2.3 Overcoming Inter-Scene Changes 141

    8.2.4 The Whole Algorithm 143

    8.2.5 Bandwidth Discussion 145

8.3 Experiments 147

    8.3.1   *Data Preparation* 147

    8.3.2   *Evaluation Metric* 148

    8.3.3   *Compared Methods* 149

    8.3.4   *Comparisons under Intra-Scene Changes* 150

    8.3.5   *Comparisons under Inter-Scene Changes* 152

    8.3.6   *Performance-Bandwidth Trade-off Analysis* 154

    8.3.7   *Ablation Study* 155

8.4 Chapter Conclusion 156

Chapter 9   Conclusions and Future Work 158

References 160

Acknowledgement 181

About the Authors 182

Chapter 1

Introduction

Nowadays, people benefit by big data and the Internet of Things (IoT). With the gradual development of information science and computer science, Artificial Intelligence (AI) has gradually become a powerful assistant to help human production and daily life. In computer vision[1,2], natural language processing[3,4], robot[5,6], recommendation system[7] and multimodal learning[8,9], etc., AI has been more powerful than human beings. The most feasible approach to AI is machine learning, which can be categorized into supervised learning, unsupervised learning, semi-supervised and reinforcement learning. Supervised machine learning adopts statistical learning to carry out likelihood learning based on large-scale collected training data, and the trained model will be applied in the test process[10]. At present, due to the development of hardware such as Graphics Processing Unit (GPU), neural network-based deep learning has been greatly developed. From AlexNet[11] to ResNet[12], and then to Transformer[13], the design of the model and algorithm is constantly improved and the effect of the model is also rising.

However, big data is not just equivalent to massive amounts of collected data. An explosion of new data is generated on the Internet every day. According to statistics, about 240,000 images are uploaded every minute on the famous social network Facebook, requiring a storage capacity of more than 450Gb. Faced with such scenarios, traditional machine learning relies on pre-collected data or the assumption that no new tasks will emerge, i.e. training in a static environment. The model trained under this premise is difficult to adapt to a changing scene. Once the environment changes, the trained model will no longer be applicable and will be quickly eliminated, which is not sustainable. For example, researchers at Harvard Medical School collected 6, 9 and 12 month patient waiting time data at Massachusetts General Hospital and trained a random forest on each of the three data points. The random forest model was used to predict the waiting time for patients over the following year. The experiments show that better test results can be obtained by training with more data. However, as time goes by, the data changes and the model effect of training on the past data keeps declining. Recollecting data to train the model would be extremely costly to solve the following difficulties:

(1) Due to privacy, storage costs and other reasons, the past data may not be available;

(2) Without considering the past data, training only the new data will make the old knowledge overwritten and forgotten;

(3) Each time new data is generated, the cost of storage and computation is too high to retrain the old and new data.

Continual learning, also known as lifelong learning and incremental learning, aims to learn a range of new knowledge over long periods of time. Human beings constantly encounter new scenarios in their living environment, which requires brains to learn new knowledge and to make sure they do not forget the past knowledge. The form of new knowledge can be new task, new category, new domain or even new modal, which are respectively called task-incremental continual learning, class-incremental continual learning, domain-incremental continual learning and modal-incremental continual learning. In recent years, continual learning has been applied to a number of artificial intelligence applications, such as medical analysis, speech recognition, flow monitoring, behavior recognition, target recognition and so on. Most continual learning depends on updating the sharing neural network parameters among sequential tasks. Updating the sharing model directly by optimization such as gradient descent will make the model forget old knowledge after learning new knowledge, which is called Catastrophic Forgetting (CF) issue. In order to solve catastrophic forgetting, the current methods are mainly divided into rehearsal-based method, regularization-based method and architecture-based method.

With the application and popularization of artificial intelligence in various fields, researchers have a strong interest in and demand for the study of this field. In this book, we introduce the core theory and some applications of continual learning. We also introduce some of our research items in continual learning. Our contributions are several-fold:

(1) We propose a cross-task margin loss function, which connects each isolation task and sets two-level angular margins in the cross-entropy to encourage the intra-class/task compactness and the inter-class/task discrepancy;

(2) We propose to evaluate the example influence in continual learning;

(3) We propose an asymmetric gradient discrepancy metric for parallel continual learning;

(4) We propose an augmented graph neural network for multi-label continual learning;

(5) We propose a federated continual learning based on Fisher information fusion;

(6) We propose a centroid-based rehearsal selection strategy to sample more representative samples in rehearsal-based continual learning and present a centroid distance distillation algorithm to reduce forgetting;

(7) We propose to apply continual learning to V2X perception application.

1.1 Static and Dynamic Artificial Intelligence

Since the deep learning algorithm was proposed in 2006, the application of artificial intelligence technology has made a breakthrough.

Since 2012, the explosive growth of data has provided sufficient "nourishment" for artificial intelligence, and deep learning algorithms have achieved breakthroughs in speech and visual recognition. The level of artificial intelligence is built on the basis of machine learning. In addition to advanced algorithms and hardware computing power, big data is the key to machine learning. Big data can help train machines and improve their intelligence. The richer and more complete the data, the more accurate the machine identification will be, so big data will be the real capital for enterprises to compete. Big data is the fuel for the progress of artificial intelligence and an important basis for the building of artificial intelligence. Through the learning of large amounts of data, the judgment and processing ability of machines will continue to improve, and the level of intelligence will also continue to improve.

Machine learning is a science that studies how to use computers to simulate or realize human learning activities. It is one of the most intelligent and cutting-edge research fields in artificial intelligence. Since the 1980s, machine learning, as a way to achieve artificial intelligence, has aroused wide interest in the field of artificial intelligence. Especially in the last decade, the research work in the field of machine learning has developed rapidly, and it has become one of the important topics in artificial intelligence. Machine learning is not only applied in knowledge-based systems, but also in many fields such as natural language understanding, non-monotone reasoning, machine vision, pattern recognition and so on. Whether a system has the ability to learn has become a sign of whether it has "intelligence". Machine learning research is mainly divided into two research directions: the first is the traditional machine learning research, which mainly studies the learning mechanism, focusing on exploring the learning mechanism of simulating human; the second is the research on machine learning under the environment of big data, which mainly studies how to effectively use information and pays attention to obtaining hidden, effective and understandable knowledge from huge amounts of data.

After 70 years of tortuous development, machine learning, represented by deep learning, has made breakthroughs in many aspects by drawing on the multi-layered structure of the human brain, the layer-by-layer analysis and processing mechanism of the connection and interaction of neurons, and the powerful parallel information processing capability of self-adaptation and self-learning, among which the most representative is the field of image recognition. In terms of application of machine learning, with the development of the fifth-generation mobile communication technology, the communication between devices will have higher bandwidth and lower delay, which gives birth to more artificial intelligence applications, such as autonomous driving, VR and so on. It clears the way for these technologies to be implemented and applied.

The commonly used machine learning methods are supervised learning and unsupervised learning. A simple generalization is supervised by whether an input data has a label. If the input data has labels, it is supervised learning, while no label is unsupervised learning. Supervised learning is a kind of learning algorithm when the correct output of exponential data set is known. Because inputs and outputs are known, it means that there is a relationship between inputs and outputs, and supervised learning algorithms are designed to discover and summarize this "relationship". Common supervisory algorithms include linear regression, neural network, decision tree, support vector machine, KNN, etc. Unsupervised learning refers to a class of learning algorithms for unlabeled data. Because there is no label information, this means that patterns or structures need to be discovered and summarized from the data set. Unsupervised algorithms include principal component analysis (PCA), anomaly detection method, self-coding algorithm, deep belief network, Herbie learning method, generative adversarial network, etc.

In the research of intelligent technology and application system of the autonomous unmanned system (such as automatic driving, smart factory and other application fields), the sample category mode and attribute space are constantly changing, and the model is required to have high robustness to the dynamic environment. The continual learning model with high plasticity, re-growth and dynamics, which can select and evolve models and memorize historical knowledge effectively, is an important theoretical and technical support for an intelligent learning system to adapt to a dynamic environment. Static AI has a fixed and immutable model structure, which is suitable for the case of a given amount of data and a fixed category. However, in the real world in the era of big data, data and categories are always increasing, and the fixed structure model cannot adapt to it. Therefore, dynamic artificial intelligence is needed to deal with the ever-increasing data flow.

1.2 Theory of Continual Learning

Continual learning (CL), also known as incremental learning, is based on the idea of constantly learning from the outside world in order to achieve the autonomous, incremental development of more complex skills and knowledge. A continual learning system can be viewed as an adaptive algorithm capable of learning from a continual flow of information that becomes progressively available over time and does not predefine the number of tasks to be learned. Continual learning aims to continually learn new knowledge from a sequence of tasks over a lifelong time. In contrast to traditional supervised learning, the continual setting helps machine learning work like a more realistic human learning by acquiring a new skill quickly with new training data.

In recent years, with the continuous development of information technology, there is an explosive growth of various data. Traditional machine learning algorithms can only achieve good performance when the distribution of test data is similar to that of training data. In other words, they cannot continuously and adaptively learn in a dynamic environment. This ability of adaptive learning is a characteristic of any intelligent system.

Deep neural networks have shown the best learning ability in many applications. However, learning with incremental updates to data using this method suffers from catastrophic interference or forgetting problems, causing the model to forget how to solve old tasks after learning new ones. Continual learning (CL) alleviates this problem. Continual learning is to simulate the brain learning process. It learns the independently and identically distributed (IID) stream data in a certain order and then updates the model incrementally according to the execution results of the task. The significance of continual learning is to efficiently transform and use the knowledge already learned to complete the learning of new tasks, and can greatly reduce the problem caused by forgetting. The study of continual learning is of great significance for intelligent computing systems to adaptively adapt to environmental changes.

1.2.1 Scenarios of Continual Learning

As shown in Table 1.1, for different incremental types, continual learning can be categorized into three scenarios[14]: Task-incremental (Task-IL), Domain-incremental (Domain-IL) and Class-incremental (Class-IL) continual learning.

Table 1.1   Three scenarios of continual learning

Type	Example
Task-IL	With task given, is it the 1st or 2nd class? (e.g., 0 or 1)
Domain-IL	With task unknown, is it a 1st or 2nd class? (e.g., in [0,2,4,6,8] or in [1,3,5,7,9])
Class-IL	With task unknown, which digit is it? (i.e. choice from 0 to 9)

The Task-incremental continual learning solves sequential tasks with task-ID provided. The model always knows which tasks need to be performed. This is the simplest continual learning solution. Since the task identity is always provided, in this case you can use task-specific components to train the model. The typical network architecture used in this case has "multiple" output layers, meaning that each task has its own output unit, but the rest of the network is shared between tasks.

Domain-incremental continual learning solves sequential tasks without task-ID, the task identity is not available at the time of testing. However, the model only needs to address the current task. They don't have to infer which task it is. A typical example of this is when the structure of the protocol task is always the same, but the input distribution changes.

Class-incremental continual learning solves sequential tasks by inferring task-ID, the models must be able to solve every task they have seen so far and infer the tasks they present. As shown in Fig. 1.1, we call this case Class-incremental Learning because it includes the common real-world problem of learning a new class step by step. It is an algorithm that learns continuously from a sequential data stream in which new classes occur. At any time, the learner is able to perform multi-class classification for all classes observed so far. At the very least, a visual object classification system should be able to incrementally learn about new classes, when training data for them becomes available. As illustrated in iCaRL[15], Class-IL contains three elements: (1) it should be trainable from a stream of data in which examples of different classes occur at different times; (2) it should at any time provide a competitive multi-class classifier for the classes observed so far; (3) its computational requirements and memory footprint should remain bounded, or at least grow very slowly, with respect to the number of classes seen so far. The first two criteria express the essence of Class-IL. The third criterion prevents trivial algorithms, such as storing all training examples and retraining an ordinary multi-class classifier whenever new data becomes available.

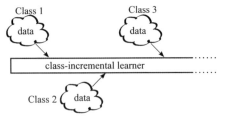

Fig. 1.1 Class-incremental learning

An interesting setting is a scenario where there are no clear task boundaries in the data stream, this is called the blurry setting. For each task in a dataset, the blurry setting keeps the majority of the examples while we randomly swap a small percentage of the examples with other tasks. A larger swap percentage corresponds to more blurry boundaries. In Aljundi, Lin, Goujaud, et al.[16], they keep 90% of the data for each task, and introduce 10% of data from the other tasks. They make comparisons to the other studied selection methods. Since tasks are not disjoint, forgetting is not as severe as completing disjoint tasks. In Bang, Kim, Yoo, et al.[17], the disjoint continual learning setup exaggerates the catastrophic forgetting since it never exposes seen classes in successive tasks, but it is deviated from the real-world where new classes do not show up exclusively. Conversely, the blurry setup makes the task boundaries faint in a way that each task contains a small number of classes also present in the other tasks.

1.2.2 Challenges of Continual Learning

All the while, because the trained task is unavailable, catastrophic forgetting[18] is the main challenge for continual learning. It happens when the learner forgets the knowledge of old tasks while learning a new task. As the model incrementally learns new knowledge, old knowledge is overwritten and gets a drop in performance. When the model learns more new tasks, it will forget more about the old task. When AI is learning the current Task B, its knowledge of the previous Task A will suddenly be lost. Especially when the past data is not visible, the AI model keeps learning new classes and will cover past knowledge. Thus, in continual learning, the learning model requires incremental construction and dynamic updating of internal representations as the distribution of tasks changes dynamically over its lifetime. Ideally, one part of the model's internal representation will be generic (immutable) enough to be reused for similar tasks, while another part should be dynamically adapted to accommodate the new task-specific representation.

1.2.3 Approaches of Continual Learning

The state-of-art methods for continual learning can be categorized into three main branches to solve the catastrophic forgetting problem.

First, the regularization-based methods[18-22], this line of work introduces additional regularization terms in the loss function to consolidate previous knowledge when learning new tasks. These methods are based on regularizing the parameters corresponding to the old tasks, penalizing the feature drift on the old tasks and avoiding storing raw inputs. As shown in Fig. 1.2, Kirkpatrick et al.[18] limits changes to parameters based on their significance to the previous tasks using Fisher information; LwF[23] is a data-focused method, and it leverages the knowledge distillation combined with a standard cross-entropy loss to mitigate forgetting and transfer knowledge by storing the previous parameters. Thuseethan et al.[24] propose an indicator loss, which is associated with the distillation mechanism that preserves the existing upcoming emotion knowledge. Yang et al.[25] introduce an attentive feature distillation approach to mitigate catastrophic forgetting while accounting for semantic spatial- and channel-level dependencies. The regularization-based procedures can protect privacy better because they do not collect samples from the original dataset. This kind of method can be divided into two kinds: data-based and prior-based. The basic idea of the data-based method is to take the old task model as the expert model, to conduct knowledge distillation of the expert model, and to guide the new task model, namely the student model training. The basic idea of the priority-based method is to estimate the distribution of model parameters, which can be used as priors in the learning of new tasks, so as to inhibit catastrophic forgetting. Usually, the prior-based method will estimate the importance of all neural network parameters, assuming that these parameters are independent, which can ensure the feasibility of its theory. When training new missions, there will be a penalty for changing important parameters of old missions.

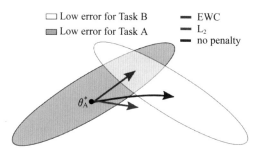

Fig. 1.2　Elastic weight consolidation

Second, the rehearsal-based methods[26-33], which sample a limited subset of data from the previous tasks or a generative model as the memory. The stored memory is replayed while learning new tasks to mitigate forgetting. In ER[34], this memory is retrained as the extended training dataset during the current training; RM[17] is a replay method for the blurry setting; iCaRL[15] selects and stores samples closest to the feature mean of each class for replaying; A-GEM[35] resets the training gradient by combining the gradient on the memory and training data; Ye et al.[36] propose a Teacher-Student network framework. The Teacher module would remind the Student about the information learnt in the past. Continual learning methods based on rehearsal can be divided into three types: joint training, gradient change and data generation. (1) Joint training. In this method, the model is trained on a limited subset of stored samples when training new tasks. The performance upper bound of such methods is to combine training with all data from the old task and data from the new task. (2) Gradient change. Because the stored sample set is relatively small, the overfitting problem is easy to occur in the training process. Using the idea of Constrained Optimization, the method directly operates on the gradient of old and new tasks to suppress forgetting. (3) Data generation. With the old task sample unavailable, the Pseudo Rehearsal method is used to generate the old task sample for joint training.

Third, the parameter isolation based methods[37-40], which generate task-specific parameter expansion or sub-branch. When no limits apply to the size of networks. Expert Gate[37] grows new branches for new tasks by dedicating a model copy to each task. Packnet[38] iteratively assigns parameter subsets to consecutive tasks by constituting binary masks.

1.3 Content of This Book

This book aims to enable readers to have a more comprehensive understanding of the theory and application of continual artificial intelligence. This book is divided into 9 chapters.

- Chapter 1, Introduction. This chapter introduces the theory and applications of continual learning in general and is used to provide sufficient knowledge base of the proposed method.
- Chapter 2, Multi-Domain Multi-Task Rehearsal for Continual Learning. This chapter introduces one of the main reasons for catastrophic forgetting, the continual domain drift, and proposes a novel multi-domain multi-task rehearsal method.
- Chapter 3, Exploring Example Influence in Continual Learning. This chapter introduces how to evaluate the example influence in continual learning, and proposes to make use the example influence in improving the training.
- Chapter 4, Measuring Asymmetric Gradient Discrepancy in Parallel Continual Learning. This chapter defines the parallel continual learning, and proposes to evaluate the gradient conflict with an asymmetric metric.
- Chapter 5, Multi-Label Continual Learning Using Augmented Graph Convolutional Network. This chapter introduces the first application of continual learning in multi-task classification and proposes an augmented graph neural network in building class relations.
- Chapter 6, Towards Long-Term Remembering for Federated Continual Learning. This chapter introduces the second application of continual learning in federated learning and proposes a Fisher

accumulation mechanism.

- Chapter 7, Centroid-based Rehearsal in Online Continual Learning. This chapter introduces to address catastrophic forgetting from the perspective of continuous domain drift and propose to solve forgetting via centroid distance distillation.
- Chapter 8, Dynamic V2X Perception from Road-to-Vehicle Vision. Based on the above theory, an adaptive road-to-vehicle perception method is proposed to establish a V2X perception framework based on road-to-vehicle visual information.
- Chapter 9, Conclusions and Future Work. This chapter summarizes the work of this book and presents several possible future directions of continual learning.

Chapter 2

Multi-Domain Multi-Task Rehearsal for Continual Learning

Rehearsal, seeking to remind the model by storing old knowledge in continual learning, is one of the most effective ways to mitigate catastrophic forgetting, i.e. biased forgetting of previous knowledge when moving to new tasks. However, the old tasks of the most previous rehearsal-based methods suffer from the unpredictable domain shift when training the new task. This is because these methods always ignore two significant factors. First, the Data Imbalance between the new task and the old tasks that makes the domain of old tasks prone to shift. Second, the Task Isolation among all tasks will make the domain shift toward unpredictable directions; To address the unpredictable domain shift, in this chapter, we propose Multi-Domain Multi-Task (MDMT) rehearsal to train the old tasks and the new task parallelly and equally to break the isolation among tasks. Specifically, a two-level angular margin loss is proposed to encourage the intra-class/task compactness and inter-class/task discrepancy, which keeps the model from domain chaos. In addition, to further address domain shift of the old tasks, we propose an optional episodic distillation loss on the memory to anchor the knowledge for each old task. Experiments on benchmark datasets validate that the proposed approach can effectively mitigate the unpredictable domain shift.

2.1 Introduction

Continual learning, also known as continual learning and incremental learning, aims to continually learn new knowledge from a sequence of tasks over a lifelong time. In contrast to traditional supervised learning, the lifelong setting helps machine learning work like a more realistic human learning by acquiring a new skill quickly with new training data. All the while, catastrophic forgetting[18,41] is the main challenge for continual learning, which happens when the learner forgets the knowledge of old tasks while learning a new task. To seek a balance between the old tasks and the new task, many methods have been proposed to handle the catastrophic forgetting in recent years. Following Lange, Aljundi, Masana, et al.[42], their methods can be categorized into Rehearsal[35,43,44], Regularization[18,23,35] and Parameter Isolation[38,45]. Regularization-based and parameter isolation-based methods store no data from old tasks and highly rely on extra regularizers or architectures, resulting in their lower performance than the rehearsal-based methods. Rehearsal-based methods store a small number of samples in the training set, the model will retrain the saved data when training the new task to avoid forgetting.

At each step of continual learning, the most existing rehearsal-based methods[15,16,35,46] focus on training the new task while treating the stored data from old tasks as the constraints to preserve their performance. However, the old tasks in these methods may suffer from unpredictable domain shift that arises from two significant factors in the continual learning process: (1) The Data Imbalance between the old tasks and the new task. The shrinkage of training data of old tasks makes the domains prone to shift and leads to the catastrophic forgetting issue. (2) The Task Isolation among all tasks (old and new), which makes such domain shift toward unpredictable directions and the boundary between any two tasks may become weak.

To address the unpredictable domain shift, in this chapter, we propose a Multi-Domain Multi-Task (MDMT) Rehearsal method inspired by the multi-domain multi-task learning[47] that considers both multiple tasks and multiple domains, and trains them equally. The procedure can be seen in Fig. 2.1. Specifically, we first retrain the old tasks along with new task training parallelly rather than setting them as the constraints. We separate all these tasks by a Cross-Domain Softmax, which extends the Softmax for each isolated task by combining the logits of all other seen tasks and separates them from each other. Then, to further alleviate the unpredictable domain shift, we propose to leverage a Two-level Angular Margin (TAM) loss to encourage the intra-class/task compactness and the inter-class/task discrepancy on the basis of Cross-Domain Softmax. In addition, we present an optional Episodic Distillation (ED) loss on all buffer memories for old tasks that suppress the domain shift by storing the latent representations of each sample in memories. We evaluate our MDMT rehearsal on four popular continual learning datasets for image classification and achieve new state-of-the-art performance. The experimental results show the proposed MDMT rehearsal can significantly mitigate the unpredictable domain shift. Our contributions are three-fold: (1) We propose a Multi-Domain Multi-Task Rehearsal method for continual learning, which parallelly and equally trains the old and new tasks and separates them by a Cross-Domain Softmax function. (2) We propose a Two-level Angular Margin (TAM) loss for continual learning to further boost the Cross-Domain Softmax for the sake of intra-class/task compactness and the inter-class/task discrepancy. (3) We build an optional Episodic Distillation loss to reduce the domain shift in lifelong progress.

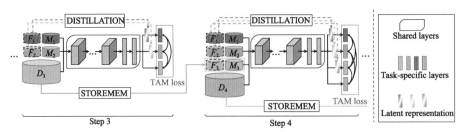

Fig. 2.1 Training procedure of the proposed MDMT rehearsal based continual learning

2.2 Methodology

2.2.1 Multi-Domain Multi-Task Rehearsal

Suppose there are T different tasks with respect to datasets $\{\mathcal{D}_1, \cdots, \mathcal{D}_T\}$. For the t-th dataset (task), $\mathcal{D}_t = \{(x_{t,1}, y_{t,1}), \cdots, (x_{t,N_t}, y_{t,N_t})\}$, where $x_{t,i} \in \mathcal{X}_t$ is the i-th input data, $y_{t,i} \in \mathcal{Y}_t$ is the corresponding label and N_t is the number of samples. $\mathcal{D}_t$ can be split into a training set $\mathcal{D}_t^{\text{trn}}$ and a testing set $\mathcal{D}_t^{\text{tst}}$, and we denote $\mathcal{D}_t$ as $\mathcal{D}_t^{\text{trn}}$ in our presentation for simple denotation. Continual learning aims at learning a predictor $f_t: \mathcal{X}_k \to \mathcal{Y}_k$, $k \in \{1, \cdots, t\}$, which can predict tasks that have been learned at any time. The rehearsal-based continual learning[15,35,46,48,49] builds a memory buffer $\mathcal{M}_k \subset \mathcal{D}_k$ with small-size for each previous task k, i.e. $|\mathcal{M}_k| \ll |\mathcal{D}_k|$. Following Chaudhry, Gordo, Dokania, et al.[50], when training a task $t \in \{1, \cdots, T\}$, for all $\mathcal{M}_k$ that $k < t$, the rehearsal-based continual learning can be modeled as a single objective optimizing problem:

$$\begin{aligned}\arg\min_{\theta, \theta_t} \quad & l(f_\theta, f_{\theta_t}, \mathcal{D}_t), \\ \text{s.t.} \quad & l(f_\theta, f_{\theta_k}, \mathcal{M}_k) \leq l(f_\theta^{t-1}, f_{\theta_k}^{t-1}, \mathcal{M}_k), \forall k < t,\end{aligned} \quad (2.1)$$

where l is the empirical loss. θ is the shared parameter across all tasks while θ_k and θ_t are the task-specific parameters. The constraints above are designed to prevent the performance degradation of previous tasks. Then, the problem can be reduced to find an optimal gradient that benefits all tasks. To inspect the increase in old tasks' loss, Chaudhry, Ranzato, Rohrbach, et al.[35], Guo, Liu, Yang, et al.[46], Chaudhry, Gordo, Dokania, et al.[50] compute the angle between the gradient of each old task and the proposed gradient update on the current task.

However, such a single objective optimization on the current task for rehearsal-based continual learning over-emphasizes the new task while ignoring the difference among tasks. In other words, the old tasks can only play the role of source domain to be transferred into the current training model. The domain of old tasks will significantly shift because of the rectified gradient that the gradient norm of new task is much larger than the old tasks', which may induce the domain overlap.

In contrast, this chapter treats the problem as a Multi-Domain Multi-Task (MDMT) learning problem to jointly and equally improve the current task as well as the old tasks:

$$\underset{\theta,\{\theta_1,\cdots,\theta_t\}}{\arg} \{\min l(f_\theta, f_{\theta_t}, \mathcal{D}_t), \min l(f_\theta, f_{\theta_k}, \mathcal{M}_k), \cdots, \min l(f_\theta, f_{\theta_1}, \mathcal{M}_1)\},$$

$$\text{s.t.} \quad d(f_i, f_j) \geq d(f_i^{t-1}, f_j^{t-1}), i, j \in [1, t], i \neq j, \tag{2.2}$$

where $f_i = f_\theta(\mathcal{D}_i)$ if $i = t$ and $f_i = f_\theta(\mathcal{M}_i)$ if $i < t$. d means the distance between two domains. For the t tasks w.r.t. datasets $\{\mathcal{D}_1, \cdots, \mathcal{D}_t\}$, a MDMT rehearsal model trains t tasks parallelly and equally. The constraints above mean the domain distance between any two tasks should not be smaller than the model trained on the last task. Note that we only consider the situation that the tasks are irrelevant as the common continual learning.

We make two key operations to solve the Eq. (2.2) efficiently. First, we transform the multi-objective optimization as a single-objective optimization problem by ensembling all these objectives as the traditional solution to multi-task learning[51,52].

$$\arg\min_\theta l(f_\theta, f_{\theta_t}, \mathcal{D}_t) + \sum_{k=1}^{t-1} l(f_\theta, f_{\theta_k}, \mathcal{M}_k). \tag{2.3}$$

Second, there exists high memory-cost to calculate the distance between any two domains and store old predictors f_θ^{t-1}, but we can do this in a simple yet effective way by extending the Softmax function for each task as

$$l_k = -\frac{1}{N_k} \sum_{n=1}^{N_k} \log \frac{e^{(W_{y_n}^k)^T x_n + b_{y_n}}}{\sigma_n}, \tag{2.4}$$

where

$$\sigma_n = \sum_{j=1}^{C_k} e^{(W_j^k)^T x_n + b_j} + \sum_{i=1, i \neq k}^{t} \sum_{j=1}^{C_i} e^{(W_j^i)^T x_n + b_j}. \tag{2.5}$$

N_k is the batch size for task k and $W_j^k \in \mathbb{R}^d$ denotes the j-th column of the weight $W^k \in \mathbb{R}^{d \times C_k}$ in the last fully-connected layer for task k and C_k is the class number. We name this extension as Cross-Domain Softmax (CDS), which combines the logits from other classifiers and is similar to a native Softmax to a classification problem with total $\sum_{k=1}^{t} C_k$ class. Here, we discuss the difference. For MDMT rehearsal, different tasks never share a same classifiers as common classification, i.e. the classifiers for different tasks lack mutual perception. By combining the logits from other tasks, the tasks can perceive and separate from each other. The previous methods update the model by the optimal gradient that highly rely on the angle between the gradients of old and new tasks. In contrast, we directly obtain the hybrid gradient for the shared layers by ensembling the gradients from the new task and the old tasks as $\tilde{g} \leftarrow \sum_{k=1}^{t} g_k$.

We compare our MDMT rehearsal with several well-known rehearsal-based lifelong works:

- iCaRL[15] saves small number of samples to make the model not forget old class, but they classify samples by the nearest prototype, which is not suitable for task-incremental continual learning because the task-specific parameters are ignored.
- GEM/A-GEM[35,53] proposes to solve forgetting by finding the optimal gradient that saves the old tasks from being corrupt, and it focuses on training the new task with single objective optimization while ignoring the domain shift of old tasks.
- ER[35] extends Experience Replay[34] for reinforcement continual learning and is proven better than A-GEM. However, it never consider the relations among all tasks, which makes the domains of old tasks shift significantly.
- PRD[54] proposes to treat continual learning as a multi-task learning problem and proposes to build a distillation module with one saved CNN expert as the teacher for each old task. Differently, we would like to build a MDMT rehearsal that leverages the expanded

Softmax without saving many extra models.

2.2.2 Two-Level Angular Margin Loss

The proposed MDMT rehearsal helps to jointly and equally train the new task and retrain the old tasks, making all tasks perceive each other. Nonetheless, the Softmax loss is not efficient enough because it does not explicitly encourage intra-class compactness and inter-class discrepancy, in coping with which, large margin based Softmax is widely used in recent discriminative problems[55,56]. However, these methods cannot be directly applied to MDMT rehearsal based continual learning because these methods place the large margin only to single task and can not be applied to multiple tasks scenario.

In this chapter, we propose two levels margin, i.e. class level and task level, on Softmax for each task [Eq. (2.4)]. Our work is based on the popular large margin based Softmax method Arcface[55] where the large margin is added to the angle between weight and feature, which has been proven effective and efficient. Specifically, Arcface deletes the bias and transforms the logit fed into the Softmax as $W_j^T x_i = \|W_j\| \|x_i\| \cos \theta_j$ where θ_j is the angle between the weight W_j and the feature x_i, then an angular margin m is placed between different classes

$$l = -\frac{1}{N} \sum_{i=1}^{N} \log \frac{e^{s \cdot \cos(\theta_{y_i} + m)}}{e^{s \cdot \cos(\theta_{y_i} + m)} + \sum_{j=1, j \neq y_i}^{n} e^{s \cdot \cos \theta_j}}, \quad (2.6)$$

where the individual weight $\|W_j\|$ is fixed to 1 by l_2 normalization and the embedding feature $\|x_i\|$ is fixed to s by l_2 normalization and rescale. The normalization on features and weights makes the predictions only depend on the angle between them. Such a geodesic distance margin between the sample and centers makes the prediction gain more intra-class compactness and inter-class discrepancy.

Based on Eq. (2.6), we propose our Two-level Angular Margin (TAM) loss for the task $k \in [1, t]$

$$l_k = -\frac{1}{N_k} \sum_{n=1}^{N_k} \log \frac{e^{s \cdot \cos[(\theta^k_{y_n} + m^c) + m^t]}}{\sigma_n}, \quad (2.7)$$

where

$$\sigma_n = e^{s \cdot \cos[(\theta^k_{y_i} + m^c) + m^t]} + \sum_{j=1, j \neq y_i}^{C_k} e^{s \cdot \cos(\theta^k_j + m^t)} + \sum_{i=1, i \neq k}^{t} \sum_{j=1}^{C_i} e^{s \cdot \cos \theta^i_j}. \quad (2.8)$$

In Eq. (2.7), we add class-level margin m^c and task-level margin m^t on the angular. m^c is similar to m in Eq. (2.6), which controls the intra-task class compactness and discrepancy[55]. m^t controls the task compactness and discrepancy, which ensures the knowledge of each task not to mix up with others.

As shown in Fig. 2.2, the proposed TAM loss produces two advantages for MDMT rehearsal based continual learning. First, TAM

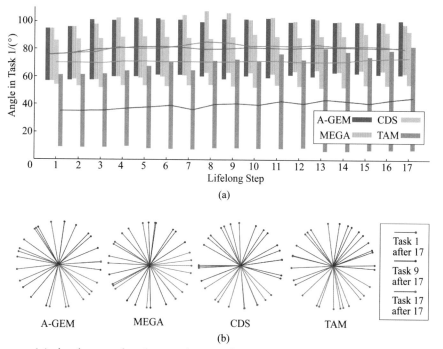

(a) the changes of angle range between feature and the target weight center of Task 1 along the continual learning. (b) the angulars relations of class centers of Tasks 1, 9 and 17 after trained on Task 17.

Fig. 2.2 On Permuted MNIST

helps the model to better discriminate into a task. Although the CDS has a better angle between feature and its target weight, TAM loss even reduce the angle to one smaller than CDS, which expresses the effect of m^c. Second, TAM loss mitigates the domain overlap caused by the domain shift by forcing tasks to separate. We can also see that for the angles among weights center, TAM loss can significantly separate old and new tasks, which expresses the effect of m^t. However, it is still difficult to omit the domain shift because of the extreme data imbalance between old tasks and the new task. Thus, we construct an optional Episodic distillation loss for the MDMT rehearsal based lifelong process.

2.2.3 Episodic Distillation

In this chapter, we propose a simple yet effective solution to further mitigate the domain shift for old tasks named Episodic Distillation (ED) loss. The main role the ED loss played is to reduce the feature distribution change along with the lifelong process as far as possible. First, apart from the sampled training data stored in memory, i.e. $\mathcal{M}_k = \{(x_{k,1}, y_{k,1}), \cdots, (x_{k,|M_k|}, y_{k,|M_k|})\} \subset \mathcal{D}_k$, we also store the corresponding latent representations when they are first trained, denoted as $\mathcal{F}_k = \{f_{k,1}, \cdots, f_{k,|M_k|}\}$. Then, we train the model with an updated objective:

$$\arg\min_{\theta} l(f_\theta, f_{\theta_t}, \mathcal{D}_t) + \sum_{k=1}^{t-1}[l(f_\theta, f_{\theta_k}, \mathcal{M}_k) + \tilde{l}(f_\theta, \mathcal{F}_k)], \quad (2.9)$$

where

$$\tilde{l}(f_\theta, \mathcal{F}_k) \triangleq \frac{1}{N_k} \sum_i \tilde{l}_i(f_\theta(x_{k,i}), f_{k,i}). \quad (2.10)$$

$\tilde{l}_i$ is the ED loss that can be in many formats, and we choose the Mean Square Error (MSE). By training with Eq. (2.9) in each step, we can ease the shift effectively.

ED loss is an optional loss function and builds extra memory buffers to save the latent representation for each sample in memories. The extra memory buffers do increase the memory cost to some extent, but the size is still very small in compared with the whole training set. In our implementation, we save the representation from the fc layer before the last one, which is a vector with length from 256 to 2,048 for different network. That means the cost of the representation memory is even smaller than the data memory.

2.2.4 Total Algorithm

We follow A-GEM[35] that unites memory of all old tasks for efficient training. Let $\mathcal{M} = U_{k<t} \mathcal{M}_k$ and $\mathcal{F} = U_{k<t} \mathcal{F}_k$ be the united data and representation memory for old tasks. For each step, we will sample a batch of data from the united memory. In this way, the previous tasks will be optimized by an average gradient instead of all gradients for previous tasks, which speeds up the training.

Our algorithm includes training and evaluation procedures. First, the storage of memory feature in StoreMem to be as the anchor of old task in current task training. Second, the gradient to be updated depends not only on old and current tasks using TAM loss, but also on the feature difference by ED loss. The evaluation procedure is similar with the previous works.

2.3 Experiments

2.3.1 Experimental Settings

We evaluate the proposed method on four image recognition datasets. (1) Permuted MNIST[18]: this is a variant of standard MNIST dataset of handwritten digits with 20 tasks. Each task has a fixed random permutation of the input pixels which is applied to all the images of that task. (2) Split CIFAR[57]: this dataset consists of 20 disjoint subsets of CIFAR-100 dataset[57], where each subset is formed by randomly sampling 5 classes without replacement from the original 100 classes. (3) Split CUB[18]: the CUB dataset[58] is split into 20 disjoint subsets by randomly sampling 10 classes without replacement from the original 200 classes. (4) Split

AWA[18]: this dataset consists of 20 subsets of the AWA dataset[15]. Each subset is constructed by sampling 5 classes with replacement from a total of 50 classes and the same class can appear in different subsets.

We leverage four existing metrics to evaluate the performance and catastrophic forgetting. (1) Average Accuracy ($A_t \in [0,1]$) after the model has been trained continuously done till Task $t \in \{1, \cdots, T\}$. In particular, A_T is the average accuracy on all the tasks after the last task has been learned. (2) Forgetting Measure[59] ($F_t \in [-1, 1]$) is the average forgetting after the model has been trained continuously with all the mini-batches for Task $t \in \{1, \cdots, T\}$. (3) Learning Curve Area[59] (LCA) ($LCA \in [0,1]$) is the area of the convergence curve for any average b-shot performance after the model has been trained for all the T tasks, where $b \in [0, \beta]$. (4) Long-Term Remembering[46] (LTR) ($LTR \geqslant 0$) quantifies the accuracy drop on each task relative to the accuracy just right after the task has been learned. The detailed descriptions and the formulas can be shown in the supplementary materials.

Following the previous works[35,46,53], for Permuted MNIST, we adopt a standard fully-connected network with two hidden layers, where each layer has 256 units with ReLU activation. For Split CIFAR, we use a reduced ResNet18[12]. For Split CUB and Split AWA, we use a standard ResNet18.

2.3.2 Comparison with the State-of-the-arts

We compare the proposed method with the state-of-the-art methods including EWC[18], MAS[42], RWalk[59], PI[57], GEM[53], MER[60], ER[35], A-GEM[35] and MEGA[46]. Specifically, EWC, MAS, RWalk and PI are regularization-based methods that prevent the important weights from changing too much. GEM, MER, ER, A-GEM and MEGA are rehearsal-based methods that rectifies the gradient guided by the stored data. VAN is a single supervised model trained continuously on the sequence of tasks. We also compare with the baseline that jointly trains all datasets with different classifiers together.

First, as shown in Table 2.1, the quantitative results of the proposed method outperform other state-of-the-arts. For A_T, the performances of our method show the superiority on all four datasets. This indicates the less forgetting on old tasks and better learning on new tasks through the lifelong training by reducing unpredictable domain shift. F_T evaluates the fine-grained batch-level forgetting on all tasks and never cares the Acc value. We get good F_T except on Split CIFAR with slight worse (0.04 vs. 0.03) than MER. MER has a better F_T but poor A_T because it adopts a complex meta learning strategy. For LCA_{10}, it evaluates the training speed on the first 10 training batches for each task, our method has the best LCA_{10} only on Split CUB. This is because the TAM and ED losses may slow the early training to mitigate domain overlap, but the following training will be improved significantly. LTR focuses on long-term remembering and our method outperforms other methods on these datasets except Split CUB. We think this is because the dataset CUB contains similar classes of birds, which means less impact of TAM and ED losses because of similar representations. In Fig. 2.3, we show the average accuracy trends in the continual process (from A_1 to A_T), which also indicate the better performance of the MDMT-R.

Table 2.1 Comparison with different state-of-the-arts

Method	Permuted MNIST				Split CIFAR			
	$A_T/\%$	F_T	LCA_{10}	LTR	$A_T/\%$	F_T	LCA_{10}	LTR
Joint	95.30	—	—	—	68.30	—	—	—
VAN	47.55± 2.37	0.52± 0.026	0.259± 0.005	5.375± 0.194	40.44± 1.02	0.27± 0.006	0.309± 0.011	2.613± 0.174
EWC	68.68± 0.98	0.28± 0.010	0.276± 0.002	3.292± 0.135	42.67± 4.24	0.26± 0.039	0.336± 0.010	2.493± 0.427
MAS	70.30± 1.67	0.26± 0.018	0.298± 0.006	—	42.35± 3.52	0.26± 0.030	0.332± 0.010	—
RWalk	85.60± 0.71	0.08± 0.007	0.319± 0.003	—	42.11± 3.69	0.27± 0.032	0.334± 0.012	—

continued

Method	Permuted MNIST				Split CIFAR			
	$A_T/\%$	F_T	LCA_{10}	LTR	$A_T/\%$	F_T	LCA_{10}	LTR
MER	—	—	—	—	37.27±1.68	0.03±0.030	0.051±0.101	—
GEM	89.50±0.48	0.06±0.004	0.230±0.005	—	61.20±0.78	0.06±0.007	0.360±0.007	—
A-GEM	89.32±0.46	0.07±0.004	0.277±0.008	0.716±0.048	61.28±1.88	0.09±0.018	0.350±0.013	0.643±0.124
ER	90.47±0.14	0.03±0.001	0.184±0.004	0.367±0.013	63.97±1.30	0.06±0.006	0.349±0.105	0.451±0.333
MEGA	91.21±0.10	0.05±0.001	0.283±0.004	0.524±0.017	66.12±1.94	0.06±0.015	0.375±0.012	0.356±0.114
MDMT-R	94.33±0.04	0.02±0.000	0.298±0.003	0.247±0.009	69.20±1.60	0.04±0.010	0.334±0.008	0.283±0.099

Method	Split CUB				Split AWA			
	$A_T/\%$	F_T	LCA_{10}	LTR	$A_T/\%$	F_T	LCA_{10}	LTR
Joint	65.60	—	—	—	64.80	—	—	—
VAN	53.89±2.00	0.13±0.020	0.292±0.008	0.976±0.215	30.35±2.81	0.04±0.013	0.214±0.008	0.202±0.090
EWC	53.56±1.67	0.14±0.024	0.292±0.009	1.021±0.210	33.43±3.07	0.08±0.021	0.257±0.011	0.675±0.214
MAS	54.12±1.72	0.13±0.013	0.293±0.008	—	33.83±2.99	0.08±0.022	0.257±0.011	—
RWalk	54.11±1.71	0.13±0.013	0.293±0.009	—	33.63±2.64	0.08±0.023	0.258±0.011	—
PI	55.04±3.05	0.12±0.026	0.292±0.010	—	33.86±2.77	0.08±0.022	0.259±0.011	—
A-GEM	61.82±3.72	0.08±0.021	0.302±0.011	0.456±0.174	44.95±2.97	0.05±0.014	0.287±0.012	0.178±0.082
ER	73.63±0.52	0.01±0.005	0.265±0.004	0.001±0.001	54.27±4.05	0.02±0.030	0.293±0.009	0.014±0.015
MEGA	80.58±1.94	0.01±0.017	0.311±0.010	0.002±0.002	54.28±4.84	0.05±0.040	0.305±0.015	0.070±0.114
MDMT-R	84.27±1.63	0.01±0.015	0.337±0.013	0.017±0.014	61.56±3.36	0.02±0.027	0.298±0.008	0.002±0.002

Note: The numbers are averaged across 5 runs using a different seed each time.

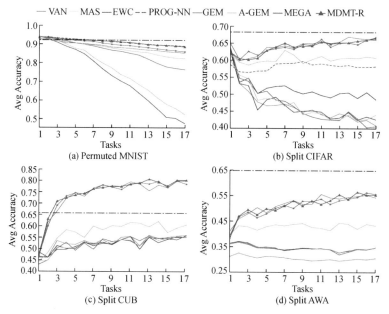

Fig. 2.3 Average accuracy trend (from A_1 to A_T) on four datasets in the lifelong process

In Table 2.2, we then analyze the importance of the main components including TAM and ED losses on Split CIFAR. The first row is the results with only vanilla Softmax. By adding ED loss, the average accuracy gets a little improvement. By adding TAM loss, the performance obtains larger gains, and we select the best m_t and m_c as the hyperparameters where $m_t = 0$ and $m_c = 0$ means the Cross-Domain Softmax. By adding both TAM loss and ED loss, we obtain a dramatic improvement in performance compared to the vanilla Softmax and the state-of-the-art methods, which means the TAM and ED losses can significantly reduce the forgetting.

Table 2.2 Ablation study on split CIFAR

m^t	m^c	ED	$A_T / \%$	F_T	LCA_{10}	LTR
—	—	—	65.44±1.13	0.052±0.006	0.371±0.008	0.377±0.076
—	—	✓	66.44±2.22	0.050±0.009	0.370±0.014	0.307±0.066
0.0	0.0	—	67.15±2.02	0.053±0.012	0.353±0.006	0.411±0.097
0.1	0.0	—	67.49±1.55	0.049±0.010	0.354±0.005	0.369±0.096
0.0	0.01	—	67.45±1.09	0.059±0.008	0.354±0.005	0.483±0.064
0.1	0.01	—	67.68±1.72	0.052±0.008	0.350±0.007	0.390±0.062

continued

m^t	m^c	ED	$A_T/\%$	F_T	LCA_{10}	LTR
0.4	0.01	—	67.28±0.97	0.053±0.012	0.347±0.006	0.394±0.106
0.4	0.05	—	66.68±1.23	0.063±0.005	0.333±0.005	0.473±0.077
0.4	0.1	—	64.97±1.13	0.084±0.009	0.324±0.008	0.680±0.086
0.1	0.01	✓	68.64±1.35	0.059±0.016	0.334±0.008	0.297±0.103

2.3.3 Domain Shift Observation

In this section, we would like to show some observations of domain shift using t-distributed Stochastic Neighbor Embedding (t-SNE)[61] on Permuted MNIST. First, in order to intuitively reflect the task relation of the proposed method during the training process, we visualize the final feature distribution, i.e. trained after the Task 17, of Tasks 1, 9 and 17 in Fig. 2.4. A-GEM and MEGA cannot guarantee the task boundaries, which means generating some mix area and makes the task easy to misclassify. The proposed MDMT rehearsal separates each class in three tasks while

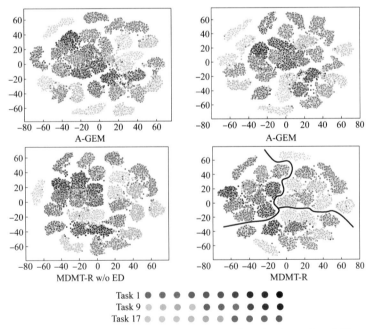

Fig. 2.4 Final t-SNE of the features extracted from Tasks 1, 9 and 17 on Permuted MNIST after the training on Task 17

obtain explicit task boundary, which means the proposed method is able to encourage the intra-class/task compactness and inter-class/task discrepancy. As shown in Fig. 2.5, we also show the domain shift of Task 1 after the model trained on Tasks 1, 9 and 17, respectively. The previous methods A-GEM and MEGA cannot reduce the domain shift at all, which makes them sustainable to forget. The proposed MDMT rehearsal method can significantly mitigate the unpredictable domain shift. Without ED loss, our MDMT rehearsal still gets some unpredictable domain shift (such as Task 1 after 1 and 9) because of the shrink of training data.

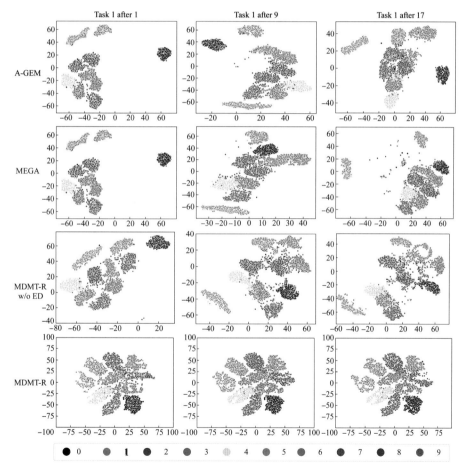

Fig. 2.5　t-SNE of the features from Task 1 on Permuted MNIST after the continual learning on Tasks 1, 9 and 17

2.4 Chapter Conclusion

In this chapter, we address catastrophic forgetting, a major drawback of state-of-the-art lifelong learning study, by considering the unpredictable domain shift of old tasks in the training sequence. To this end, we proposed a Multi-Domain Multi-Task rehearsal method, which effectively makes all tasks perceive each other. Then we proposed a Two-level Angular Margin loss to further encourage the intra-class/task compactness and inter-class/task discrepancy. Finally, an optional Episodic Distillation loss was proposed to mitigate domain shift. We have tested the proposed approach on four image classification benchmark datasets. Extensive experiments show the superiority of our approach over state-of-the-art methods.

Chapter 3

Exploring Example Influence in Continual Learning

3.1 Introduction

By mimicking human-like learning, Continual Learning (CL) aims to enable a model to continuously learn from novel knowledge (new tasks, new classes, etc.) in a sequential order. The major challenge in CL is to harness catastrophic forgetting and knowledge transition, namely the Stability-Plasticity dilemma, with Stability (S) showing the ability to prevent performance drops for old tasks and Plasticity (P) referring if the new task can be learned rapidly and unimpededly. Intuitively speaking, a robust CL system should achieve outstanding S and P through sequential learning.

The sequential paradigm means CL does not access past training data. Compared to traditional machine learning, the training data in CL is thus more precious. It is valuable to explore the influence difference on S and P among training examples. Following the accredited influence chain "Data-Model-Performance", exploring this difference is equivalent to tracing from performance back to example difference. With appropriate control, this may improve the learning pattern towards better SP. On top of this, the goal of this chapter is to explore the reasonable influence from each training example to SP, and apply the example influence to CL training.

To understand example influence, one classic successful technique is the Influence Function (IF)[62], which leverages the derivation chain rule from a test objective to training examples. However, directly applying the chain rule leads to computing the inverse of Hessian with the complexity of $O(nq^2+q^3)$ (n is the number of examples and q is parameter size), which is computationally intensive and may out-of-memory in neural networks. In this chapter, we propose a novel meta-learning algorithm, called MetaSP, to compute example influence via simulating IF. We design on the basis of the rehearsal-based CL framework, which avoids forgetting via retraining a part of old data. First, a pseudo update is held with example-level perturbations. Then, two validation sets sampled from seen data are used to compute the gradients on example perturbations. The gradients are regarded as the example influence on S and P. As shown in Fig. 3.1(a), examples can be distinguished by the value of influence on S and P.

To leverage the two independent kinds of influence in CL, we need to take full account of the influence on both S and P. However, the influence on S and P may interfere with each other, which leads us to make a trade-off. This can be seen as a Dual-Objective Optimization (DOO) problem, which aims to find solutions not dominated (no other better solution) by any other one, i.e. Pareto optimal solutions[63]. The solutions can be treated as the example influence on SP. Following the gradient-based MGDA algorithm[64], we obtain the fused example influence on SP by meeting the Karush-Kuhn-Tucker (KKT) condition, as illustrated in Figure 3.1(b).

Finally, we show the fused influence can be used to control the update of model and optimize the storage of rehearsal in Figure 3.1(c). On one hand, the fused influence can be directly used to control the magnitude of training loss for each example. On the other hand, under a fixed memory budget, the fused influence can be used to select appropriate examples storing and dropping, which keeps the rehearsal memory always have larger positive influence on SP.

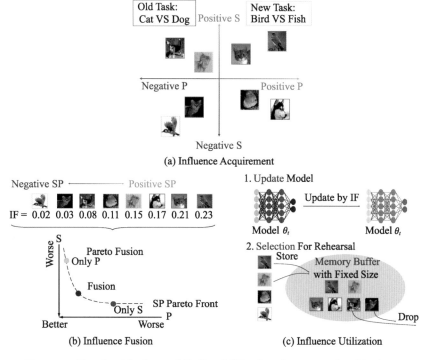

Given an old task with classes "Cat" and "Dog" and a new task with classes "Bird" and "Fish", we compute the influence on S and P for each example. Then, we fuse the two kinds of influence towards SP Pareto front. We also show that example influence can be used to adjust model update and optimize rehearsal selection.

Fig. 3.1 Training examples have different influences on Stability and Plasticity

In summary, our contributions are four-fold: (1) Inspired by the influence function, we study CL from the perspective of example difference and propose MetaSP to compute the example influence on S and P. (2) We propose to trade off S and P influence via solving a DOO problem and fuse them towards SP Pareto optimal. (3) We leverage the fused influence to control model update and optimize the storage of rehearsal. (4) The verification contribution: by considering the example influence, in our experiments on both task- and class-incremental CL, better S and more stable P can be observed.

3.2 Methodology

3.2.1 Preliminary: Rehearsal-based CL

Given T different tasks w. r. t. datasets $\{\mathcal{D}_1, \cdots, \mathcal{D}_T\}$, Continual Learning (CL) seeks to learn them in sequence. For the t-th dataset (task), $\mathcal{D}_t = \{(x_t^{(n)}, y_t^{(n)})\}_{n=1}^{N_t}$ is split into a training set $\mathcal{D}_t^{\text{trn}}$ and a test set $\mathcal{D}_t^{\text{tst}}$, where N_t is the number of examples. At any time, CL aims at learning a multi-task/multi-class predictor to predict tasks/classes that have been learned (say task-incremental and class-incremental CL). To suppress the catastrophic forgetting, the rehearsal-based CL[15,35,46,53,60] builds a small size memory buffer $\mathcal{M}_t$ sampled from $\mathcal{D}_t^{\text{trn}}$ for each task (i.e. $|\mathcal{M}_t| \ll |\mathcal{D}_t^{\text{trn}}|$). At training phase, the data in the whole memory $\mathcal{M} = \bigcup_{k<t} \mathcal{M}_k$ will be retrained together with the current tasks. Accordingly, a mini-batch training step of Task t in rehearsal-based CL is denoted as

$$\min_{\theta_t} l(\mathcal{B}_{\text{old}} \cup \mathcal{B}_{\text{new}}, \theta_t), \quad \mathcal{B}_{\text{old}} \subset \mathcal{M} \text{ and } \mathcal{B}_{\text{new}} \subset \mathcal{D}_t^{\text{trn}}, \tag{3.1}$$

where l is the empirical loss. θ_t is the trainable parameters at Task t and is updated from scratch.

3.2.2 Example Influence on Stability and Plasticity

The S of a task is evaluated by the performance difference on the test set after training on any later tasks, which is also known as Forgetting[35]. The P of a task is defined as the ability to integrate new knowledge, which is regarded as the test performance of this task. As many existing CL methods demonstrate, the SP inevitably interferes mutually.

Lemma 3.1 (Example Influence on SP)

At the training on the t-th task, with a sampled example $x^{\text{trn}} \in \mathcal{D}_t^{\text{trn}}$, the example influence from x^{trn} to Stability S_t^k and Plasticity P_t for $k < t$ can be evaluated by the gap from deleting it then retraining the model:

$$I_S(\mathcal{D}_k^{\text{tst}}, x^{\text{trn}}) = p(\mathcal{D}_k^{\text{tst}} | \theta_{t-1}, \mathcal{D}_t^{\text{trn}}) - p\left(\mathcal{D}_k^{\text{tst}} \middle| \theta_{t-1}, \frac{\mathcal{D}_t^{\text{trn}}}{x^{\text{trn}}}\right),$$

$$I_P(\mathcal{D}_t^{\text{tst}}, x^{\text{trn}}) = p(\mathcal{D}_t^{\text{tst}} | \theta_{t-1}, \mathcal{D}_t^{\text{trn}}) - p\left(\mathcal{D}_t^{\text{tst}} | \theta_{t-1}, \frac{\mathcal{D}_t^{\text{trn}}}{x^{\text{trn}}}\right),$$

where $\mathcal{D}_t^{\text{trn}}/x^{\text{trn}}$ denotes the dataset $\mathcal{D}_t^{\text{trn}}$ without the training example x^{trn}.

However, deleting every example to compute full influences is impractical due to the highly computational cost. Instead, the performance change can be indicated by the loss change, which leads to a derivable way to approximate the influence:

$$I_S(\mathcal{D}_k^{\text{tst}}, x^{\text{trn}}) \stackrel{\text{def}}{=\!=} \frac{\partial l(\mathcal{D}_k^{\text{tst}})}{\partial \varepsilon}, \quad I_P(\mathcal{D}_t^{\text{tst}}, x^{\text{trn}}) \stackrel{\text{def}}{=\!=} \frac{\partial l(\mathcal{D}_t^{\text{tst}})}{\partial \varepsilon}, \quad (3.2)$$

where ε is the weight perturbation to the training example and $\stackrel{\text{def}}{=\!=}$ means define. This influence can be computed by the Influence Function[62] that will be introduced in the next section.

3.3 Meta Learning on Stability and Plasticity

3.3.1 Influence Function for SP

A mini-batch, $\mathcal{B}$, from the training data is sampled, and the normal model update is

$$\hat{\theta} = \arg\min_{\theta} l(\mathcal{B}, \theta). \quad (3.3)$$

In Influence Function (IF)[62], a small weight perturbation ε is added to the training example $x^{\text{trn}} \in \mathcal{B}$

$$\hat{\theta}_{\varepsilon, x} = \arg\min_{\theta} l(\mathcal{B}, \theta) + \varepsilon l(x^{\text{trn}}, \theta), x^{\text{trn}} \in \mathcal{B}. \quad (3.4)$$

We can easily promote this to the mini-batch

$$\hat{\theta}_{E, \mathcal{B}} = \arg\min_{\theta} l(\mathcal{B}, \theta) + E^{\mathrm{T}} L(\mathcal{B}, \theta), \quad (3.5)$$

where L denotes the loss vector for a mini-batch and $E \in \mathbb{R}^{|\mathcal{B}| \times 1}$ denotes the perturbation on each example in it. It is easy to know that the example influence $I(\mathcal{D}^{\text{tst}}, \mathcal{B})$ is reflected in the derivative $\nabla_E l(\mathcal{D}^{\text{tst}}, \hat{\theta}_{E, x})|_{E=0}$. By the chain rule, the example influence in IF can be computed by

$$I(\mathcal{D}^{\text{tst}}, \mathcal{B}) \stackrel{\text{def}}{=\!=} \nabla_E l(\mathcal{D}^{\text{tst}}, \hat{\theta}_{E, x})|_{E=0} = -\nabla_\theta l(\mathcal{D}^{\text{tst}}, \hat{\theta}) H^{-1} \nabla_\theta^{\mathrm{T}} L(\mathcal{B}, \hat{\theta}), \quad (3.6)$$

where $\boldsymbol{H} = \nabla_\theta^2 l(\mathcal{B}, \hat{\theta})$ is a Hessian. Unfortunately, the inverse of Hessian requires the complexity $O(|\mathcal{B}|q^2 + q^3)$ and huge storage for neural networks (maybe out-of-memory), which is challenging for efficient training.

In Eq. (3.6), we have $I(\mathcal{D}^{tst}, \mathcal{B}) = [I(\mathcal{D}^{tst}, x^{trn}) | x^{trn} \in \mathcal{B}]$ and find the loss will get larger if $I(\mathcal{D}^{tst}, x^{trn}) > 0$, which means the negative influence on the test set $\mathcal{D}^{tst}$. Similarly, $I(\mathcal{D}^{tst}, x^{trn}) < 0$ means the positive influence on the test set $\mathcal{D}^{tst}$. Fortunately, the second-order derivation in IF is not necessary under the popular meta learning paradigm such as Hospedales, Antoniou, Micaelli, et al.[65], instead we can easily get the derivation like IF through a one-step pseudo update. In the following, we will introduce a simple yet effective meta-based method, named MetaSP, to simulate IF at each step with a two-level optimization to avoid computing Hessian inverse.

3.3.2 Simulating IF for SP

Based on the meta learning paradigm, we transform the example influence computation into solving a meta gradient descent problem, named MetaSP. For each training step in a rehearsal-based CL, we have two mini-batches data $\mathcal{B}_{old}$ and $\mathcal{B}_{new}$ in respect to old and new tasks. Our goal is to obtain the influence on S and P from every example in $\mathcal{B}_{old} \cup \mathcal{B}_{new}$. Note that both S-aware and P-aware influence are applied to every example regardless of old or new tasks. That is, the contribution of an example is not deterministic. Data of old tasks may also affect the new task in positive, and vice-versa. In rehearsal-based CL, we turn to computing the derivations $\nabla_E l(\mathcal{V}_{old}, \hat{\theta})|_{E=0}$ for example influence.

To compute the derivation, as shown in Fig. 3.2, our MetaSP has two key steps:

(1) Pseudo update. This step is to simulate Eq. (3.5) in IF via a pseudo update

$$\hat{\theta}_{E,\mathcal{A}} = \arg\min_\theta l(\mathcal{B}_{old} \cup \mathcal{B}_{new}, \theta) + \boldsymbol{E}^\top \boldsymbol{L}(\mathcal{B}_{old} \cup \mathcal{B}_{new}, \theta), \quad (3.7)$$

Exploring Example Influence in Continual Learning 37

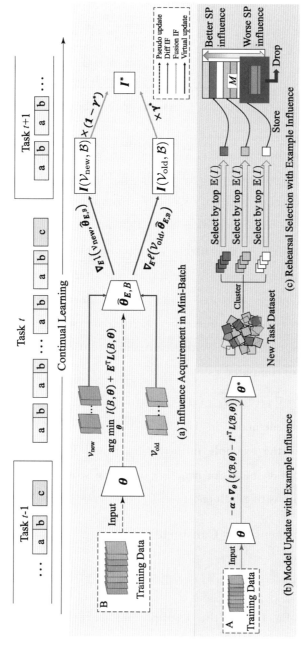

Fig. 3.2 Evaluating and making use of example influence in mini-batch Continual Learning

(a) At each iteration in CL training, MetaSP updates in pseudo and use two validation sets representing old tasks and new task to obtain the example influence on S and P. The two kinds of influence are fused towards a Pareto optimal. (b) The computed influence can be directly used to update CL model and (c) select examples for rehearsal storing and dropping.

Algorithm 1: Computation of Example Influence(MetaSP)	
Input: $\mathcal{B}_{old}, \mathcal{B}_{new}, \mathcal{V}_{old}, \mathcal{V}_{new}$;	// Training batches, Validation batches
Output: $\boldsymbol{I}^*$;	// Pareto example influence on SP
1. $\hat{\theta}_{E,\mathcal{B}} = \arg\min_\theta l(\mathcal{B}_{old} \cup \mathcal{B}_{new}, \theta) + \boldsymbol{E}^T\boldsymbol{L}(\mathcal{B}_{old} \cup \mathcal{B}_{new}, \theta)$	// Pseudo update
2. $\boldsymbol{I}(\mathcal{V}_{old}, \mathcal{B}) = \nabla_{\boldsymbol{E}} l(\mathcal{V}_{old}, \hat{\theta}_{E,\mathcal{B}})$;	// Gradient from old val loss
3. $\boldsymbol{I}(\mathcal{V}_{new}, \mathcal{B}) = \nabla_{\boldsymbol{E}} l(\mathcal{V}_{new}, \hat{\theta}_{E,\mathcal{B}})$;	// Gradient from old val loss
4. $\gamma^* \leftarrow$ Eq.(3.11);	// Optimal fusion hyper-parameter
5. $\boldsymbol{I}^* = \gamma^* \cdot \boldsymbol{I}(\mathcal{V}_{old}, \mathcal{B}) + (1-\gamma^*) \cdot \boldsymbol{I}(\mathcal{V}_{new}, \mathcal{B})$	// Influence fusion

where $\boldsymbol{L}$ denotes the loss vector for a mini-batch combining both old and new tasks.

(2) Compute example influence. This step computes example influence on S and P for all training examples as simulating Eq. (3.6). Based on the the pseudo updated model in Eq. (3.7), we compute S- and P-aware example influence via two validation sets $\mathcal{V}_{old}$ and $\mathcal{V}_{new}$. Noteworthily, because the test set $\mathcal{D}^{tst}$ is unavailable at training phase, we use two dynamic validation sets $\mathcal{V}_{old}$ and $\mathcal{V}_{new}$ to act as the alternative in the CL training process. One is sampled from the memory buffer ($\mathcal{V}_{old}$) representing the old tasks, and the other is from the seen training data representing the new task ($\mathcal{V}_{new}$). With $\boldsymbol{E}$ initialized to $\boldsymbol{0}$, the two kinds of example influence are computed as

$$\boldsymbol{I}(\mathcal{V}_{old}, \mathcal{B}) = \nabla_{\boldsymbol{E}} l(\mathcal{V}_{old}, \hat{\theta}_{E,\mathcal{B}}), \boldsymbol{I}(\mathcal{V}_{new}, \mathcal{B}) = \nabla_{\boldsymbol{E}} l(\mathcal{V}_{new}, \hat{\theta}_{E,\mathcal{B}}). \quad (3.8)$$

Generally, each elements in two influence vectors $\boldsymbol{I}(\mathcal{V}_{old}, \mathcal{B})$ and $\boldsymbol{I}(\mathcal{V}_{new}, \mathcal{B})$ represents the example influence on S and P. Similar to IF, elements with positive value mean negative influence while elements with negative value mean positive influence.

3.4 Using Influence for Continual Learning

3.4.1 Before Using: Influence for SP Pareto Optimality

As shown in Eq. (3.8), the example influence is equal to the derivation from validation losses of old and new tasks to the perturbations $\boldsymbol{E}$. However, the two kinds of influence are independent and interfere with each other. That is, using only one of them may fail the other performance. We prefer to find a solution that makes a trade-off between the influence on both S and P.

Thus, we integrate the two influences $I(\mathcal{V}_{old}, \mathcal{B})$ and $I(\mathcal{V}_{new}, \mathcal{B})$ into a DOO problem with two gradients from different objectives.

$$\min_{E}\{l(\mathcal{V}_{old}, \hat{\theta}_{E,\mathcal{B}}), l(\mathcal{V}_{new}, \hat{\theta}_{E,\mathcal{B}})\}. \tag{3.9}$$

The goal of the problem in Eq.(3.9) is to obtain a fused way that satisfies the SP Pareto optimality.

Lemma 3.2 (SP Pareto Optimality)

(1) (Pareto Dominate) Let E_a, E_b be two solutions for Problem (3.9), E_a is said to dominate E_b ($E_a \prec E_b$) if and only if $l(\mathcal{V}, \hat{\theta}_{E_a,\mathcal{B}}) \leqslant l(\mathcal{V}, \hat{\theta}_{E_b,\mathcal{B}})$, $\forall \mathcal{V} \in \{\mathcal{V}_{old}, \mathcal{V}_{new}\}$, and $l(\mathcal{V}, \hat{\theta}_{E_a,\mathcal{B}}) < l(\mathcal{V}, \hat{\theta}_{E_b,\mathcal{B}})$, $\exists \mathcal{V} \in \{\mathcal{V}_{old}, \mathcal{V}_{new}\}$.

(2) (SP Pareto Optimal) E is called SP Pareto optimal if no other solution can have better values in $l(\mathcal{V}_{old}, \hat{\theta}_{E,\mathcal{B}})$ and $l(\mathcal{V}_{new}, \hat{\theta}_{E,\mathcal{B}})$.

Inspired by the Multiple-Gradient Descent Algorithm (MGDA)[64], we transform the problem in Eq. (3.9) to a min-norm problem. Specifically, according to the KKT conditions[66], we have

$$\gamma^* = \arg\min_{\gamma} \| \gamma \nabla_E l(\mathcal{V}_{old}, \hat{\theta}_{E,\mathcal{B}}) + (1-\gamma) \nabla_E l(\mathcal{V}_{new}, \hat{\theta}_{E,\mathcal{B}}) \|_2^2, s.t., 0 \leqslant \gamma \leqslant 1. \tag{3.10}$$

Referring to the study from Sener et al.[52], the optimal γ^* is easily computed as

$$\gamma^* = \min\left(\max\left(\frac{(\nabla_E l(\mathcal{V}_{new}, \hat{\theta}_{E,\mathcal{B}}) - \nabla_E l(\mathcal{V}_{old}, \hat{\theta}_{E,\mathcal{B}}))^T \nabla_E l(\mathcal{V}_{new}, \hat{\theta}_{E,\mathcal{B}})}{\| \nabla_E l(\mathcal{V}_{new}, \hat{\theta}_{E,\mathcal{B}}) - \nabla_E l(\mathcal{V}_{old}, \hat{\theta}_{E,\mathcal{B}}) \|_2^2}, 0\right), 1\right). \tag{3.11}$$

Thus, the SP Pareto influence of the training batch can be computed by

$$I^* = \gamma^* \cdot I(\mathcal{V}_{old}, \mathcal{B}) + (1-\gamma^*) \cdot I(\mathcal{V}_{new}, \mathcal{B}). \tag{3.12}$$

This process can be seen in Fig. 3.2(a). Different from the S-aware and P-aware influence, the integrated influence consider the Pareto optimum to both S and P, i.e. reducing the negative influence on S or P and keeping the positive influence on both S and P. Then we will introduce how to leverage example influence in CL training, our algorithm can be seen in Alg. 1.

3.4.2 Model Update Using Example Influence

With the computed example influence in each mini-batch, we can easily control the model update of this mini-batch to adjust the training towards an ensemble positive direction. Given parameter θ from the previous iteration the step size α, the model can be updated in traditional SGD as $\theta^* = \theta - \alpha \cdot \nabla_\theta (l(\mathcal{B}, \theta))$, where $\mathcal{B} = \mathcal{B}_{old} \cup \mathcal{B}_{new}$. By regularizing the update with the example influence $\boldsymbol{I}^*$, we have

$$\theta^* = \theta - \alpha \cdot \nabla_\theta (l(\mathcal{B}, \theta) + (-\boldsymbol{I}^*)^T \boldsymbol{L}(\mathcal{B}, \theta)). \quad (3.13)$$

Algorithm 2: Using Example Influence in Rehearsal-based Continual Learning

Input: Initialized θ_0, Learning rate α, Training set $\{\mathcal{D}_1^{trn}, \cdots, \mathcal{D}_T^{trn}\}$, Memory $\mathcal{M}$ Output: θ_T;
// Final model

1   for   Task $t = 1 : T$ do
2       $\theta_t = $ Train New Task$(\theta_{t-1}, \mathcal{D}_t^{trn}, \mathcal{M})$
3       $\mathcal{C}_1, \mathcal{C}_2, \cdots, \mathcal{C}_{\frac{|\mathcal{M}|}{t}} \leftarrow $ K-Means$(\mathcal{D}_t^{trn})$;
4   Rank $\mathcal{C}_i$ with $\boldsymbol{E}(\boldsymbol{I}^*(x)), x \in \mathcal{C}_i$;
5   Rank $\mathcal{M}$ with $\boldsymbol{E}(\boldsymbol{I}^*(x)), x \in \mathcal{M}$;
6   for $i = 1 : \frac{|\mathcal{M}|}{t}$ do
7   Pop the bottom of $\mathcal{M}$;
8   Push the top of l_i to $\mathcal{M}$;
9   end
10 end

MetaSP offers regularized updates at every step for rehearsal-based CL, which leads the CL training to better SP but with only the complexity of $O(|\mathcal{B}|q + vq)$ (v denotes the validation size) compared with that of IF, $O(|\mathcal{B}|q^2 + q^3)$.

We show this application in Fig. 3.2(b). By updating like the above equation, we can make use of the influence of each example to a large extent. In this way, some useless examples are restrained and some positive examples are emphasized, which may improve the acquisition of new knowledge and the keeping of old knowledge simultaneously.

3.4.3 Rehearsal Selection Using Example Influence

Rehearsal in fixed budget needs to consider storing and dropping to keep the memory $\mathcal{M}$ having the core set of all old tasks. In tradition, storing and dropping are both based on randomly example selection, which ignores the influence difference on SP from each example. Given influence $I^*(x)$ representing contributions from example x to SP, we further design to use it to improve the rehearsal strategy under fixed memory budget. The above example influence on S and P is computed in mini-batch level, we can promote it to the whole dataset according to the law of large numbers, and the influence value for the example x is the value of expectation over batches, i.e. $\boldsymbol{E}(I^*(x))$.

The fixed-size memory is divided averagely by the seen task number. After task t finishes its training, we conduct our influence-aware rehearsal selection strategy as shown in Fig. 3.2(c). For storing, we first cluster all training data into $\frac{|\mathcal{M}|}{t}$ groups using K-means to diversify the store data. Each group is ranked by its SP influence value, and the most positive influence on both SP will be selected to store. For dropping, we rank again on the memory buffer via their influence value, and drop the most negative $\frac{|\mathcal{M}|}{t}$ example. In this way, $\mathcal{M}$ always stores diverse examples with positive SP influence.

3.5 Experiments

3.5.1 Datasets and Implementation Details

We use three commonly used benchmarks for evaluation: (1) Split CIFAR-10[57] consists of 5 tasks, with 2 distinct classes each and 5,000 exemplars per class, deriving from the CIFAR-10 dataset; (2) Split CIFAR-100[57] splits the original CIFAR-100 dataset into 10 disjoint subsets, each of which is considered as a separate task with 10 classes; (3) Split Mini-Imagenet[67] is a subset of 100 classes from ImageNet[68], rescaled to 32×32. Each class has 600 samples, randomly subdivided into training (80%) and test sets (20%). Mini-Imagenet dataset is equally divided into 5 disjoint tasks.

We employ ResNet-18[12] as the backbone which is trained from scratch. We use Stochastic Gradient Descent (SGD) optimizer and set the batch size 32 unchanged in order to guarantee an equal number of updates. Also, the rehearsal batch sampled from memory buffer is set to 32. We construct the SP validation sets in MetaSP by randomly sampling 10% of the seen data and 10% of the memory buffer at each training step. We set other hyper-settings following ER tricks[69], including 50 total epochs and hyper-parameters. All results are averaged over 5 fixed seeds for fairness.

To better evaluate the CL process, we suggest evaluating SP with three metrics as follows. We use the sign function $1(\cdot)$ to represent if the prediction of model is equal to the ground truth. (1) First Accuracy $\left(A_1 = \frac{1}{T}\sum_{t}\sum_{x_i \in \mathcal{D}_t^{tst}} 1(y_i, \theta_t(x_i))\right)$: for each task, when it is first trained done, we evaluate its testing performance immediately, which indicates the Plasticity, i.e. the capability of learning new knowledge. (2) Final Accuracy $\left(A_\infty = \frac{1}{T}\sum_{t}\sum_{x_i \in \mathcal{D}_t^{tst}} 1(y_i, \theta_T(x_i))\right)$: this metric is the final performance for each task, which indicates Stability, i.e. the capability of suppressing catastrophic forgetting. (3) Mean Average Accuracy $\left(A_m = \frac{1}{T}\sum_{t}\left(\frac{1}{t}\sum_{k \leqslant t}\sum_{x_i \in \mathcal{D}_k^{tst}} 1(y_i, \theta_t(x_i))\right)\right)$: this metric computes along CL process, indicating the SP performance after each task trained done. (4) Backward Transfer $\left(BWT = \frac{1}{T-1}\sum_{t=1}^{T-1}\sum_{(x,y) \in \mathcal{D}_t^{tst}} (1(y, \theta_T(x)) - 1(y, \theta_t(x))) = \frac{T}{T-1}(A_\infty - A_1)\right)$: this metric is the performance drop from first to final accuracy of each task.

3.5.2 Main Comparison Results

We compare our method against 8 rehearsal-based methods (including

GDUMB[70], GEM[53], A-GEM[35], HAL[50], GSS[16], MIR[73], GMED[72] and ER[35]). What's more, we also provide a lower bound that train new data directly without any forgetting avoidance strategy (Fine-tune) and an upper bound that is given by all task data through joint training (Joint).

In Table 3.1, we show the quantitative results of all compared methods and the proposed MetaSP in class-incremental and task-incremental settings. First of all, by controlling the training according to the influence on SP, the proposed MetaSP outperforms other methods on all metrics. With the memory buffer size growth, all the rehearsal-based CL get better performance, while the advantages of MetaSP are more obvious. In terms of the First Accuracy A_1, indicating the ability to learn new tasks, our method outperforms most of the other methods with a little numerical advantage. In terms of the Final Accuracy A_∞, which is used to measure the forgetting, we have an obvious improvement of an average of 3.17 for class-incremental setting and averagely 1.77 for task-incremental setting w.r.t. the second best result. This shows MetaSP can significantly keep stable learning of the new task while suppressing the catastrophic forgetting. It is because although the new tasks may have larger gradient to dominant the update for all rehearsal-based CL, our method improves the example with positive effective and restrain the negative-impact example. In terms of the Mean Average Accuracy A_m, which evaluates the SP throughout the whole CL process, our method shows its significant superiority with an average improvement of over 4.44 and 1.24 w.r.t the second best results in class-incremental and task-incremental settings. Moreover, with the proposed rehearsal selection strategy (Ours+RehSel), we have our A_∞ improved, which means the selected example has a clear ability for reducing catastrophic forgetting according to their influence. With our Rehearsal Selection (RehSel) strategy, we have an improvement of 0.77 on A_∞, but A_1 and A_m have uncertain performance. This means better memory may bring in worse task conflict.

Table 3.1　Comparisons on three datasets

Method	CIFAR-10 (Class increment)							
	buffer size 300				buffer size 500			
	A_1	A_∞	A_m	BWT	A_1	A_∞	A_m	BWT
Finetune			19.66		Joint			91.79
GDUMB[70]			36.92				44.27	
GEM[43]	93.90 ·	37.51 ·	55.43 ·	−70.48	92.76 ·	36.95 ·	57.36 ·	−69.76 ·
A-GEM[35]	96.57	20.02 ·	45.57 ·	−95.68 ·	96.56 ·	20.01 ·	46.52 ·	−95.69 ·
HAL[50]	91.30 ·	24.45 ·	46.34 ·	−83.56 ·	91.96 ·	27.94 ·	49.05 ·	−80.01 ·
MIR[71]	96.70 ·	38.53 ·	56.96 ·	−72.72 ·	96.65	42.65 ·	59.99 ·	−67.50 ·
GSS[16]	96.53	35.89 ·	54.33 ·	−75.80 ·	96.55 ·	41.96 ·	58.16 ·	−68.24 ·
GMED[72]	96.65	38.12 ·	58.92 ·	−73.16 ·	96.65	43.68 ·	62.56 ·	−66.21 ·
ER[35]	96.73	34.19 ·	53.72 ·	−78.18 ·	96.74	40.45 ·	57.69 ·	−70.36 ·
Ours	96.87	42.42	63.52	−68.05	96.82	49.16	67.88	−59.57
Ours+RehSel	96.85	43.76	63.69	−66.36	96.81	50.10	68.28	−58.38

Method	CIFAR-10 (Task increment)							
	buffer size 300				buffer size 500			
	A_1	A_∞	A_m	BWT	A_1	A_∞	A_m	BWT
Finetune			65.27		Joint			98.16
GDUMB[70]			73.22				78.06	
GEM[43]	96.62 ·	89.34 ·	92.49 ·	−9.09 ·	96.73 ·	90.42	92.93 ·	−7.88
A-GEM[35]	96.78 ·	85.52 ·	90.16 ·	−14.07 ·	96.71 ·	86.45 ·	90.90	−12.83 ·
HAL[50]	91.41 ·	79.90 ·	83.78 ·	−14.39 ·	92.03 ·	81.84 ·	84.19 ·	−12.73 ·
MIR[71]	96.76 ·	88.50 ·	90.87 ·	−10.33 ·	96.73 ·	90.63 ·	91.99 ·	−7.62
GSS[16]	96.56 ·	88.05 ·	90.60 ·	−10.63 ·	96.57 ·	90.38	92.19 ·	−7.73
GMED[72]	96.73 ·	88.91 ·	91.20 ·	−9.76 ·	96.72 ·	89.72 ·	92.10 ·	−8.75
ER[35]	96.93 ·	88.97 ·	91.12 ·	−9.95 ·	96.79 ·	90.60 ·	92.28 ·	−7.74
Ours	97.10	89.40	92.54	−9.62	97.31	90.91	93.38	−7.99
Ours+RehSel	97.11	89.91	92.66	−8.99	97.30	91.41	93.28	−7.36

continued

Method	CIFAR-100 (Class increment)							
	buffer size 500				buffer size 1,000			
	A_1	A_∞	A_m	BWT	A_1	A_∞	A_m	BWT
Finetune			9.14		Joint		71.25	
GDUMB[70]			11.11				15.75	
GEM[43]	85.28 •	15.91 •	29.38 •	−77.07	84.28 •	22.79 •	34.09 •	−68.32
A-GEM[43]	85.97	9.31 •	24.60 •	−85.18 •	85.66	9.27 •	24.67 •	−84.88 •
HAL[50]	67.33 •	8.20 •	22.72 •	−65.70 •	68.06 •	10.59 •	24.74 •	−63.86 •
MIR[71]	87.38	13.49 •	28.88 •	−82.09 •	87.39	17.56 •	32.48 •	−77.59 •
GSS[16]	86.03	14.01 •	28.00 •	−80.02 •	86.31	17.87 •	31.82 •	−76.04 •
GMED[72]	87.18	14.56 •	33.41 •	−80.68 •	87.29	18.67 •	38.69 •	−76.23 •
ER[35]	87.23	13.75 •	28.88 •	−81.64 •	87.33	17.56 •	32.45 •	−77.52 •
Ours	88.13	18.96	38.62	−76.85	87.58	24.78	45.20	−69.76
Ours+RehSel	87.81	19.28	39.23	−76.13	87.55	25.72	45.48	−68.69

Method	CIFAR-100 (Task increment)							
	buffer size 500				buffer size 1,000			
	A_1	A_∞	A_m	BWT	A_1	A_∞	A_m	BWT
Finetune			33.85		Joint		91.63	
GDUMB[70]			36.40				43.25	
GEM[43]	85.53 •	68.68 •	68.49 •	−18.72	85.24 •	73.71 •	72.59 •	−12.81
A-GEM[43]	85.97 •	55.28 •	58.23 •	−34.10 •	85.66 •	55.95 •	59.96 •	−33.01 •
HAL[50]	67.64 •	44.98 •	50.79 •	−25.17 •	68.62 •	50.07 •	54.01 •	−20.61 •
MIR[71]	87.42	66.18 •	67.43 •	−23.60 •	87.50	71.20 •	71.42 •	−18.10 •
GSS[16]	86.10 •	66.80 •	66.55 •	−21.44	86.44 •	71.98 •	71.00 •	−16.06
GMED[72]	87.30 •	68.82 •	72.66	−20.53	87.49	73.91 •	76.36	−15.10
ER[35]	87.29	66.82 •	67.56 •	−22.73	87.40	71.74 •	71.60 •	−17.40 •
Ours	88.94	70.03	74.07	−21.01	88.94	75.32	78.09	−15.14
Ours+RehSel	88.58	70.81	74.24	−19.74	89.03	76.14	78.27	14.32

continued

Method	Mini-Imagenet (Class increment)							
	buffer size 500				buffer size 1,000			
	A_1	A_∞	A_m	BWT	A_1	A_∞	A_m	BWT
	Finetune		11.12		Joint		44.39	
GDUMB[70]			6.22				7.15	
A-GEM[35]	50.06 •	10.69 •	22.29 •	−49.22	50.03 •	10.69 •	22.28 •	−49.16 •
MIR[73]	51.44	11.07 •	23.65 •	−50.46 •	51.25 •	11.32 •	24.09 •	−49.92 •
GSS[16]	51.63	11.09 •	23.62 •	−50.66 •	51.35 •	11.42 •	24.05 •	−49.91 •
GMED[72]	51.21	11.03 •	24.47 •	−50.23 •	50.87	11.73 •	25.50 •	−48.93 •
ER[35]	51.68	11.00 •	23.71 •	−50.84 •	51.41	11.35 •	24.08 •	−50.08 •
Ours	51.76	12.48	26.50	−49.10	50.91	14.43	25.47	−45.59
Ours+RehSel	51.81	12.74	26.43	−48.84	50.96	14.54	28.44	−45.52
Method	Mini-Imagenet (Task increment)							
	buffer size 500				buffer size 1,000			
	A_1	A_∞	A_m	BWT	A_1	A_∞	A_m	BWT
	Finetune		23.46		Joint		62.30	
GDUMB[70]			16.37				17.69	
A-GEM[35]	50.06 •	18.34 •	28.05 •	−39.65 •	50.03 •	18.78 •	28.12 •	−39.05 •
MIR[73]	51.47 •	29.10 •	35.20 •	−27.95 •	51.31 •	31.39 •	37.24 •	24.89 •
GSS[16]	51.64 •	28.67 •	35.22 •	−28.71 •	51.40 •	31.75 •	37.23 •	−24.56 •
GMED[72]	51.29 •	30.47 •	37.64 •	−26.02 •	51.00 •	32.85 •	39.66 •	−22.69 •
ER[35]	51.70 •	28.97 •	35.30 •	−28.40 •	51.55 •	31.59 •	37.36 •	−24.95 •
Ours	52.44	32.59	39.38	−24.82	52.27	36.25	41.59	−20.02
Ours+RehSel	51.73	34.36	40.48	−21.70	51.47	37.20	42.19	−17.83

3.5.3 Analysis of Dataset Influence on SP

In Fig. 3.3, we count the example with positive/negative influences on old task (S), new task (P), and total SP in Split-CIFAR-10. At each task after Task 2, we have 500 fixed-size memory and 10,000 new task data. We first find that most data of old tasks has a positive influence on S and a negative influence on P, while most data of new tasks has a positive influence on P and a negative influence on S. Even so, some data in both new and old tasks has the opposite influence. Then, for the total SP influence, most of memory data has positive influence. In contrast, examples of new tasks have near equal number of positive and negative SP influence. Thus, by clustering and storing examples via higher influence to rehearsal memory buffer, the old knowledge can be kept. By dividing all example influences equally into 5 groups from the minimum to the maximum, we find that most examples have mid-level influence, and server as the main part of the dataset. Also, the numbers of examples with large positive and negative influence are small, which means unique examples are few in the dataset. The observations suggest the example difference should be used to improve model training.

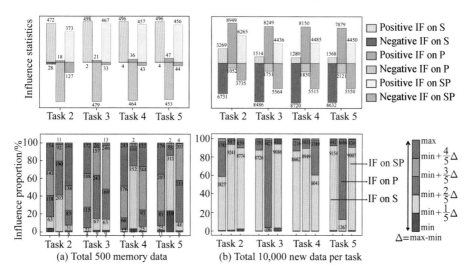

Top: Statistics of examples with positive and negative influence on S, P and SP. Bottom: We divide all example influences equally into 5 groups, and count the number in each range.

Fig. 3.3 Statistics of example influence

3.5.4 Analysis on SP Pareto Optimum

In this chapter, we propose to convert the S-aware and P-aware influence fusion into a DOO problem and use the MGDA to guarantee the fused solution is an SP Pareto optimum. We show the comparison of the First Accuracy and Final Accuracy coordinate visualization for all compared methods. We also evaluate with only Stability-aware (Ours only S) and with only Plasticity-aware (Ours only P) influence. Obviously, with only one kind of influence, our method can already get better SP than other methods. The integration of two kinds of influence yield an even more balanced SP. On the other hand, the existing methods cannot approach the SP Pareto front well.

3.5.5 Training Time

We list the training time of one-step update and total update overhead for all compared methods for Split CIFAR-10 dataset. In one-step update, we evaluate all methods with a batch on one update. Our method takes more time than other methods except for GEM, because of the pseudo update, backward on perturbation and influence fusion. To guarantee the efficiency, we utilize our proposed method only in the last 5 epochs among the total, and the rest are naive fine-tuning. The results show the strategy is as fast as other light-weight methods but achieve huge improvement on SP. We also use this setting for the comparison in Table 3.1.

3.6 Chapter Conclusion

In this chapter, we proposed to explore the example influence on Stability-Plasticity (SP) dilemma in rehearsal-based continual learning. To achieve that, we evaluated the example influence via small perturbation instead of the computationally expensive Hessian-like influence function and proposed a simple yet effective MetaSP algorithm. At each iteration in CL training, MetaSP builds a pseudo update and obtains the S-aware and P-aware example influence in batch level. Then, the two kinds of influence are combined via an SP Pareto optimal factor and can support the regular model update. Moreover, the example influence can be used to optimize rehearsal selection. The experimental results on three popular CL datasets verified the effectiveness of the proposed method. We list the limitation of the proposed method. (1) The proposed method relies on rehearsal selection, which may affect privacy and extra storage is needed. (2) The proposed method is not fast enough for online continual learning. In most situations, however, we can leverage our training tricks to reduce the time. (3) Our method is limited in the extremely small memory size. Large memory size means better remembering and an accurate validation set. The proposed method does not perform well when the memory size is extremely small.

Chapter 4

Measuring Asymmetric Gradient Discrepancy in Parallel Continual Learning

4.1 Introduction

Continual Learning (CL)[18,53,74,75], aims to continuously learn new knowledge from a sequence of tasks with non-overlapping data streams over a lifelong time. In the era of Internet of Things, people are using many smart devices, where multi-source data and tasks would be accessed at any time. A CL system should respond to parallel data streams from multiple devices. We study Parallel Continual Learning (PCL), as shown in Fig. 4.1, where an unfixed number of tasks are trained in a parallel way at any time. Specifically, according to the access time of each task, PCL builds an adaptive number of parallel data pipes, thus enabling instant response to new-coming tasks without pending.

Due to the parallel data streams from different tasks, PCL suffers from not only the catastrophic forgetting but the training conflict among parallel tasks. Most existing methods in CL are proposed to tackle the catastrophic forgetting[18,41], including regularization-based[18,37,44,57], rehearsal-based[35,46,53,76], and architecture-based[38,76,77] methods. In PCL, the training processes of different tasks are diverse, i.e. each task starts and ends training unpredictably (Fig. 4.1). Thereby the gradients from different tasks differ in direction and magnitude and may be neutralized. The gradient discrepancies lead to catastrophic forgetting and training conflict issues, which may fail the learning of some tasks. At any time in PCL, therefore, we present that the problem can be formulated to find an optimal gradient in a minimum distance multi-objective optimization, where each objective is to minimize the distance to a target gradient. In general, the distance metric is proportional to the effect of the optimal gradient on the corresponding task.

In most situations, the mentioned distance metric D between gradients is set to symmetric intuitively, such as the Euclidean distance and cosine distance. In other words, we usually have $D(x,y)=D(y,x)$ for any x and y. However, the gradient influence is imbalanced among parallel tasks in the gradient descent. For example, in Fig. 4.1, at the marked time, we have three gradients with diverse directions and magnitudes, and updating with any of them provides different influences to the other two. In the minimum distance problem, the optimal solution should have the minimum negative influence on all parallel tasks, but using symmetric metrics means the influences are optimized indistinguishably at the same time. Due to the fact that the gradients are with wide differences, the solution may have large biases, which would get the near-fitting task out of its local minimum but has less impact on a new-coming task.

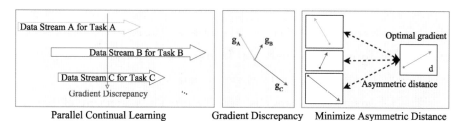

Left: PCL trains parallel tasks according to their access time without pending. Middle: At any time, gradients from different tasks (corresponding colors) have unpredicted direction and magnitude (the length of vectors). Right: We formulate PCL into a min-distance problem and propose an asymmetric distance for effective optimization.

Fig. 4.1 Overview of the proposed method in PCL

To measure the gradient discrepancy, we hold the opinion that the distance metric in the min-distance problem should be asymmetric. First, though the metric is bound up with both the gradient magnitude and direction, the influences on model training from gradients should be asymmetric, where the model should have more tolerance to small gradients even if they indicate an inverse direction. Second, because gradients are with different magnitudes, the discrepancy between two large gradients is often set to larger than that between small gradients when using symmetric distance, such as Euclidean distance. Directly optimizing using magnitude-aware distance values may lead to the solution close to large gradients and thus hinder the keeping of old tasks. To mitigate the bias from the magnitude difference, it is better to employ the magnitude ratio instead of magnitude itself.

Motivated by this, in this chapter, we propose an explicit measurement for the learning from gradient discrepancy in PCL, named Asymmetric Gradient Distance (AGD), which considers gradient magnitude ratios and directions, and sets a tolerance for smaller gradients. As shown in Fig. 4.1, the proposed AGD is used in solving the minimum distance problem with multiple gradients from parallel tasks. Then, we propose an effective optimization strategy for minimizing the gradient discrepancy to avoid self-interference. We name the strategy Maximum Discrepancy Optimization (MaxDO), which minimizes the maximum discrepancy from each gradient to the others. Moreover, to address the catastrophic forgetting issue, we follow the rehearsal strategy[53] in traditional CL and build an extra memory data stream.

The rehearsal data stream is used to provide a gradient of finished tasks in MaxDO. Solving by MaxDO with AGD, parallel training mitigates the impacts of the diverse training process and slows the catastrophic forgetting of finished tasks. Extensive results on three datasets show the superiority and effectiveness of our approach.

Our main contributions are three-fold:

(1) We formulate the PCL into a minimum distance problem and compare symmetric and asymmetric distances. We show that symmetric metrics are not effective in solving the problem and suggest asymmetric metrics.

(2) We propose an asymmetric metric, named AGD, to evaluate the gradient discrepancy, which is proportional to the gradient magnitude ratios and directions. AGD measures the imbalance of gradient influence in PCL.

(3) We propose MaxDO for minimizing gradient discrepancy of different tasks, which maximumly reduces the asymmetric discrepancy from a gradient to the others. MaxDO avoids the self-interference among gradients and reduces training conflict and catastrophic forgetting.

4.2 Methodology

4.2.1 Parallel Continual Learning

On a timeline, given a sequence of T tasks with parallel data streams $\{\mathcal{D}_1, \cdots, \mathcal{D}_T\}$ for continual training, and each data stream can be accessed and suspended at any time. For simplicity, we assume each data stream is i.i.d., and tasks are accessed in order from 1 to T and there exists no real gap that no data stream flows on the timeline. Note that traditional CL is an edge situation of PCL that all tasks are nose-to-tail. A PCL model contains a shared backbone with parameter θ to learn task-agnostic knowledge and adaptively incremental number of task-specific classifiers with parameters θ_i. When a new task is accessed, a corresponding task-specific classifier will be constructed.

In PCL, a task will be forgotten by learning any other tasks when its data stream ends. To avoid forgetting, we leverage the popular rehearsal strategy[35,46,53,76] in our training. Rehearsal builds an extra data stream sampled from all seen tasks and retrains them to suppress the forgetting of finished tasks. For convenience, we denote the rehearsal data stream as $\mathcal{D}_0$. At time t, we use $\mathcal{T}_t$ to represent the activated data streams (including $\mathcal{D}_0$). Together with the rehearsal data stream, PCL training yields the following dynamic multi-objective empirical risk minimization:

$$\min_{\theta,\{\theta_i|i\in\mathcal{T}_t\}} \{l_i(\mathcal{D}_i)|i\in\mathcal{T}_t\}. \tag{4.1}$$

Because the task-specific classifiers are updated by their own gradients $\theta_i \leftarrow \theta_i - \alpha_i \nabla_{\theta_i} l_i$ ($\forall i \in \mathcal{T}_t$) with step size α_i, we focus on the update of the shared backbone θ. At any PCL step, the goal of dynamic MOO is to optimize multiple objectives simultaneously while updating only once, and the only update of the shared parameters depends on the gradients of all in-training tasks. It will exit an uncertain number of tasks, and each task will provide a task-specific gradient on the shared parameter θ. Let $g_i = \nabla_\theta l_i$ and α be a step size for optimization. The problem of the backbone update can be formulated as follows:

$$\theta \leftarrow \theta - \alpha d^*, \text{where } d^* = f(\{g_i | \forall i \in \mathcal{T}_t\}). \tag{4.2}$$

The key question is how to compute the optimal gradient d^* via the function $f(\cdot)$. In this chapter, we define the function $f(\cdot)$ as a min-distance multi-objective problem by minimizing the gradient distance from all in-training tasks:

$$d^* = \arg\min_d \{D(d,g_i) \mid \forall i \in \mathcal{T}\}. \tag{4.3}$$

where we need to identify what distance metric D is used to measure gradient discrepancy. The motivation of Eq. (4.3) is that for the task i in PCL, its own gradient g_i is the most qualified update direction for itself. The solution d^* should be as close to every gradient as possible.

A related but different task is Federated Continual Learning (FCL)[78,79]. Most FCL methods primarily focus on serial training, involving multiple clients in training a shared task at the same time. It is meaningful to study Federated PCL in the future for the advancement of privacy protection.

4.2.2 Measuring Asymmetric Gradient Discrepancy

To measure gradient discrepancy, the Euclidean Distance (EuDist, $D(x,y)=\|x-y\| \in [0,\infty)$ and Cosine Distance (CosDist, $D(x,y)=1-\frac{x^\mathrm{T} y}{\|x\|\|y\|} \in [0,2]$) are the two most popular choices. Both of them are symmetric, i.e. $D(x,y)=D(y,x)$. A symmetric metric $D(x,y)$ means the forward influence (x to y) and backward influence (y to x) are equal. For example, given two in-training tasks A and B, the distance $D(g_A, g_B)$ represents both the effect of g_A on Task B and g_B on Task A because $D(g_A, g_B) = D(g_B, g_A)$. Note that large distance from g_A to g_B means large negative influence on the training of Task B with g_A.

However, the model update is highly related to gradient magnitude and direction, which are asymmetric to model updating. The influence of the gradient g_A on Task B may be quite different from that of the gradient g_B on Task A. In previous studies[35,53,80], the two tasks are treated as conflict when $\langle g_A, g_B \rangle < 0$. In PCL, due to the diverse training process, gradients from parallel tasks are diverse in magnitude and direction. When $\|g_A\| \ll \|g_B\|$, the gradient g_A will have little negative influence on Task B even if $\langle g_A, g_B \rangle < 0$.; when $\|g_A\| \gg \|g_B\|$ (e.g., a new Task A is accessed when Task B has been trained for some time near convergence), the update produces huge impact on Task B even if $\langle g_A, g_B \rangle > 0$. Using traditional symmetric distances can hardly represent the asymmetric update influence difference.

To effectively measure gradient discrepancy in PCL, we introduce the asymmetric metric.

Lemma 4.1 (Asymmetric Metric[81]**)**

$D: \mathcal{X} \times \mathcal{X} \to \mathbb{R}$ is an asymmetric metric (a.k.a. quasi-metric[82]) if D satisfies

(1) $D(x,y) \geqslant 0$ and $\forall x \in \mathbb{R}^d, D(x,x)=0$;

(2) $D(x,z) \leqslant D(x,y) + D(y,z), \forall x,y,z \in \mathbb{R}^d$.

The asymmetric metric does not require the symmetric property, i.e., $D(x,y)=D(y,x)$. Based on the definition, in this chapter, we design an asymmetric metric to measure gradient discrepancy named Asymmetric Gradient Distance.

Definition 4.1 Asymmetric Gradient Distance (AGD)

Given two gradient g_A and g_B, the asymmetric gradient distance is defined as

$$\hat{D}(g_A, g_B) = \begin{cases} 0, & \text{if } g_A = g_B = 0, \\ \dfrac{\|g_A - g_B\|}{\|g_B\| + \|g_A - g_B\|}, & \text{Otherwise.} \end{cases}$$

In Definition 4.1, we consider the edge situation when $g_A = g_B = 0$ to meet the definition of the asymmetric metric in Lemma 4.1. In AGD, gradient directions and magnitudes are considered. Instead of using gradient magnitude value difference, we use magnitude ratio difference to avoid the diverse training of different tasks in PCL. Therefore, we derive the corollary of the magnitude ratio:

Corollary 4.1 $\hat{D}(g_A, g_B)$ is an asymmetric metric and holds

$$\lim_{\frac{\|g_A\|}{\|g_B\|} \to \infty} \hat{D}(g_A, g_B) = 1, \quad \lim_{\frac{\|g_A\|}{\|g_B\|} \to 0} \hat{D}(g_A, g_B) = \frac{1}{2}.$$

We illustrate why AGD is qualified to evaluate the gradient discrepancy according to the definition and corollary. In Definition 4.1, we use AGD to represent the influence of g_A on Task B rather than the inverse. This is the key difference from the symmetric metrics such as Euclidean distance. Specifically, g_A may make Task B worse if $\hat{D}(g_A, g_B)$ is large (close to 1). If $\hat{D}(g_A, g_B)$ is close to 0, g_A and g_B has less conflict. Moreover, Corollary 4.1 involves that when $\|g_A\| \ll \|g_B\|$, AGD has a tolerance $\dfrac{1}{2}$ even if $\langle g_A, g_B \rangle < 0$, which means the impact of g_A on Task B is mild. This is because updating with a zero gradient will neither improve nor damage the performance. Even though, we prefer positive influence rather than non-influence. Thus, we define that the distance $\hat{D}(g_A, g_B)$ in this situation is the mid-level $\left(\frac{1}{2}\right)$ in the value range ([0,1]). See different tolerances in our experiments.

Moreover, we compare AGD [Fig. 4.2(c)] with Euclidean and cosine distance in Fig. 4.2. First, the cosine distance [Fig. 4.2(a)] is magnitude irrelevant, which ignores the magnitude difference in PCL. Second, the Euclidean distance [Fig. 4.2(b)] depends heavily on the magnitude value difference, but ignores that the gradient influence on the model update is asymmetric. For example, when $\|x\| \to 0$, EuDist will get large if we have large $\|y\|$ without any tolerance, which ignores the non-influence of zero gradient.

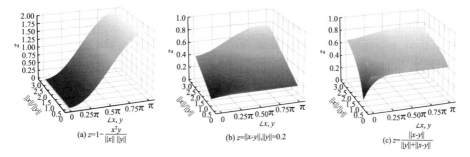

Note that the x- and y-axes are the angle (i.e. $\angle x, y$) between x and y, and the magnitude ratio $\frac{\|x\|}{\|y\|}$, respectively. (a) Cosine distance; (b) Euclidean distance where $\|y\| = 0.2$ as an example; (c) Asymmetric gradient distance.

Fig. 4.2 The measures of two gradient discrepancy from x to y

4.2.3 Maximum Discrepancy Optimization

At time t, let the optimal solution to the problem in Eq. (4.3) be d^*, where $\mathcal{T}_t$ is the index set of in-training tasks ($\mathcal{T}$ for simplicity). However, directly optimizing the problem is difficult due to the large decision space that has the same dimension as θ. Following Sener and Koltun[52], and Lin, Ye, and Zhang[83], we use linear scalarization to solve the transformed problem that allows only optimizing decision variable $w \in \mathbb{R}^{|\mathcal{T}|}$. That is, let

$d = \sum_{i \in \mathcal{T}} w_i g_i$, where $\forall w_i \geq 0$ and $\sum_{i \in \mathcal{T}} w_i = 1$, we have

$$w^* = \arg \min_w \left\{ \hat{D}\left(\sum_j w_j g_j, g_i \right) \mid \forall i \in \mathcal{T} \right\}. \tag{4.4}$$

Algorithm 3: MaxDO ( ) inPCL

Input: Parameters θ, $\theta_{1:T}$; Step sizes α, $\alpha_{1:T}$
Output: θ, $\theta_{1:T}$
1   for t in timeline do
2       $\mathcal{T}_t \leftarrow$ in-training task index;
3   for $i \in \mathcal{T}_t$ do
4          $\mathcal{B}_i \sim \mathcal{D}_i$;
5          $\theta_i \leftarrow \theta_i - \alpha_i \nabla_{\theta_i} l_i(\mathcal{B}_i)$;
6          $g_i = \nabla_\theta l_i(\mathcal{B}_i)$;
7   end
8       $W^* \leftarrow$ Optimization by Eq.(4.5);
         $d^* \leftarrow$ Final graident from Eq.(4.6);
9       $\theta \leftarrow \theta \leftarrow \alpha d^*$;
10 end

Each objective of the dual problem will be highly affected by the minimum discrepancy, i.e. each gradient itself. For example, by minimizing objective $\hat{D}\left(\sum_j w_j g_j, g_i \right)$, weight w_j is more like to be activated than others. Thus, multiple objectives will be compromised by multiple self-interference but fail to reduce the maximum discrepancy in the dual problem optimization.

As shown in Fig. 4.3, we propose Maximum Discrepancy Optimization (MaxDO) to reduce the maximum gradient discrepancy. Specifically, instead of the weight vector $W \in \mathbb{R}^{|\mathcal{T}|}$, we optimize a weight matrix $W \in \mathbb{R}^{|\mathcal{T}| \times |\mathcal{T}|}$, in which $\forall W_{ij} \geq 0$. W can be combined by a diagonal vector $\grave{w} = [W_{1,1}, \cdots, W_{|\mathcal{T}|,|\mathcal{T}|}]$ and an off-diagonal matrix $\widetilde{W} = W - \text{Diag}(\grave{w})$, where $\sum_{i \in \mathcal{T}} w_i = 1$ and $\sum_{j \in \mathcal{T}} \widetilde{W}_{ij} = 1$, $\forall i$. Thus, $\sum_{i,j \in \mathcal{T}} W_{ij} = |\mathcal{T}| + 1$ and the two weights are independent and can be optimized without disturbance as follows:

(1) $\widetilde{W}$, computed by Stochastic Gradient Descent (SGD), is used to make up the maximum gradient discrepancy. The objectives of any two rows $\widetilde{W}$

are different. For row i, to formulate the maximum discrepancy of gradient g_i, the objective is the combination of non-diagonal entries. The weighted other gradients should be with the smallest asymmetric distance to g_i. (2) $\dot{w}$ is obtained by the Multiple Gradient Descent Algorithm (MGDA)[64], which is to obtain a weighted gradient that does not damage any tasks with a min-norm optimization. The objective of MGDA is 0 and the resulting point satisfies the Karush-Kuhn-Tucker condition or the solution gives a Pareto descent direction that improves all tasks. For each off-diagonal entry of the i-th column, their sum means the effect of the gradient g_i reducing the maximum discrepancy from other gradients. MGDA is used to reduce the possible negative effect in MaxDO. On the other hand, MaxDO reduces the training failure of new tasks in MGDA. To sum up, our MaxDO with AGD can be computed by

$$W^* = \underbrace{\arg\min_{\widetilde{W}} \sum_{\forall i \in \mathcal{T}} \hat{D}(\sum_{j \neq i} \widetilde{W}_{i,j} g_j, g_i)}_{\text{SGD with Maximum Discrepancy}} + \underbrace{(\arg\min_{\widetilde{W}} \| \sum_j \dot{w}_j g_j \|)}_{[64]}. \quad (4.5)$$

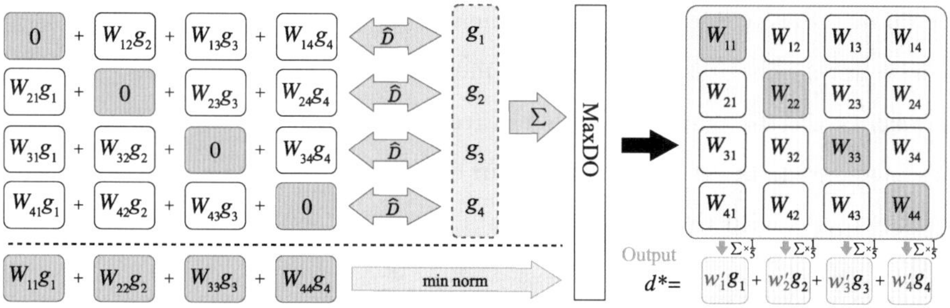

Given multiple gradients $\{g_i \mid \forall i \in \mathcal{T}\}$ ($|\mathcal{T}|=4$ for example) (1) A weight matrix W is initialized with $\frac{1}{|\mathcal{T}|}$ for each entry. (2) For each row, the off-diagonal entries are used to weighted gradients and optimized for minimum AGD to the target gradient. (3) The diagonal entries are used to optimize with min-norm with MGDA. (4) The final weight matrix is reduced by each column for the final weights (w').

Fig. 4.3 Schematic of Maximum Discrepancy Optimization

In Eq. (4.5), we can obtain an approximate solution by combining the closed-form solution and the iterative solution. Fig. 4.3 reveals the diagram of solving MaxDO. We project the solution of SGD onto the feasible set $\left(\sum_{i \neq j} W_{ij} = 1\right)$ via Softmax at each step in the multiple steps for fast convergence. First, we initialize all entries of W by $\frac{1}{|\mathcal{T}|}$. Then, the off-diagonal matrix is used to minimize the maximum gradient discrepancy via SGD and the diagonal vector is optimized by min-norm. Finally, the final weights are reduced to a vector by dividing $|\mathcal{T}|+1$ to guarantee that their sum is 1. Note that, MaxDO is implemented only when $|\mathcal{T}|>1$, i.e. multiple tasks are given at the current time. Otherwise, we have $d^* = g_1$ for the only current Task 1. Thus, the final gradient d^* is computed by

$$d^* = \begin{cases} g_1, & |\mathcal{T}|=1, \\ \sum_i \left(\frac{1}{|\mathcal{T}|+1} \sum_j W_{j,i}^*\right) g_i, & |\mathcal{T}|>1. \end{cases} \quad (4.6)$$

The detailed algorithm is shown in Algorithm 3. With the rehearsal data stream, our algorithm learns a PCL model through a timeline. At the time t on the timeline, given a mini-batch $\mathcal{B}$ from each data stream, we compute the corresponding gradients on shared and task-specific parameters. The task-specific parameters are updated directly and the gradients on the shared backbone are collected for computed the final updated gradient d. By using our MaxDO, we update the shared parameters θ with the optimal d^*.

4.3 Experiments

4.3.1 Dataset

In our experiments, three traditional image recognition datasets are transformed into parallel data streams: (1) Parallel Split EMNIST (PS-EMNIST). We split EMNIST[88] (62 classes) into 5 tasks and the size of the label set for each task, i.e. the number of classes, is set to {12,12,12, 13,13}. (2) Parallel Split CIFAR-100 (PS-CIFAR-100). We split CIFAR-100 into 10 tasks and the size of the label set for each task is set to 10.

(3) Parallel Split ImageNet TINY (PS-ImageNet-TINY). We split Tiny ImageNet[89] (200 classes) into 10 tasks, and the size of the label set for each task is set to 20. We evaluate PCL on task-incremental and class-incrementalscenarios.

All three datasets have three different label sets (three different class splits), each of which has three different timelines (when to access). For each timeline, we have three different runs with fixed seeds 1234, 1235, and 1236 for parameter initialization. In other words, we have 27 different settings for each dataset, and we report the average and standard deviation (avg±std) for each compared method in our experiments. Note that, we omit all blank time that no data stream flows for simplicity.

4.3.2 Experiment Details

We implement our experiments using Tensorflow and conduct on a single NVidia RTX 3090Ti GPU card. We take a 2-layer MLP as the backbone network for PS-EMNIST and a Resnet-18[12] for PS-CIFAR-100 and PS-ImageNet-TINY. The learning rate is set to 0.003, 0.0004 and 0.0005 for PS-EMNIST, PS-CIFAR-100 and PS-ImageNet-TINY. The SGD in MaxDO has a learning rate of 5. Each task is trained in a data stream, i.e. each data point passes only once. For each task, we set the batch size to 128 per step. For any new task in PCL, we build a new classifier, which is a fully-connected layer with a Softmax function.

To evaluate PCL, we compute the average accuracy and forgetting, following previous continual learning studies[16,35,53,72,90]. Let e_t be the end time of Task t and the final time $\bar{e} = \max(e_1, e_2, \cdots, e_T)$, the two metrics are computed as:

$$A_{\bar{e}} = \frac{1}{T}\sum_{t=1}^{T} a_{\bar{e}}^t, F_{\bar{e}} = \frac{1}{T}\sum_{t=1}^{T} a_{\bar{e}}^t - a_{e_t}^t. \qquad (4.7)$$

where a_k^j is the mean testing accuracy of task j at time k. $A_{\bar{e}}$ denotes the final average accuracy on all the tasks, and $F_{\bar{e}}$ (also known as backward transfer) means the final performance drop compared to each task that was first trained.

4.3.3 Main Results

We compare our method with MTL methods including MGDA[64], GradNorm[84], DWA[85], GradDrop[91], PCGrad[86] and RLW[83] in the PCL setting. We treat any time on the timeline as an MTL subunit to train PCL. All results of previous MTL methods are produced by ourselves with the claimed design in their methods. We also compare with some rehearsal-based SCL methods including A-GEM[35], GMED[72] and ER[35]. To adapt to the SCL methods, we merge batches from every in-training task to mimic a naive sequence learning that only a batch from the current task and a batch from the memory buffer.

We show the main comparisons with the proposed methods in Tables 4.1 and 4.2 on the three datasets. We have several major observations. First, the rehearsal strategy is useful for reducing catastrophic forgetting in PCL for all compared methods. As an extra data stream apart from in-training data streams, rehearsal provides data from the finished tasks training together with other tasks to suppress forgetting. With the rehearsal strategy, the memory may provide continual learning of finished tasks, and even better performance can be obtained, which results in positive forgetting values on F_e. Second, without task-id, class-incremental PCL has worse performance than task-incremental PCL. Surprisingly, we find that MGDA has good performance than other methods in class-incremental PCL compared with the task-incremental scenario, which means the suppressing on catastrophic forgetting is more important in class-incremental PCL. Third, the compared MTL methods are designed for balanced training and ignore the diverse training process in PCL, thus some gradients may be counteracted because of the large gradient discrepancy when updating the model. In contrast, our MaxDO with AGD obtains the best final accuracy A_e on three datasets and two scenarios, which shows our superiority in balancing plasticity and stability. For example, we have 40.770% and 9.532% for PS-ImageNet-TINY on two scenarios respectively, while the compared best value is only 39.427% and 8.318%. On one hand, the proposed AGD is used to measure the asymmetric distance between gradients to boost the effective update of each task. On the other hand, the maximum discrepancies between multiple tasks are reduced. Note that, the forgetting measure of the proposed methods may not outperform the compared methods because we got both better new tasks and final accuracy performance, their difference value may be small.

Table 4.1　Task-incremental comparisons (avg±std) on three datasets

Method (+Reherasal)	Type	PS-EMNIST		PS-CIFAR-100		PS-ImageNet-TINY	
		A_e/%	F_e/%	A_e/%	F_e/%	A_e/%	F_e/%
AvgGrad	MTL	89.344±0.231	−5.287±0.216	47.579±0.089	23.691±0.432	38.392±0.076	0.764±0.139
MGDA[64]	MTL	84.887±0.469	−4.301±0.818	48.957±0.451	25.088±0.203	38.058±0.740	1.465±0.490
GradNorm[84]	MTL	87.888±0.158	−6.197±0.183	47.210±1.323	23.474±0.688	38.226±0.769	1.647±0.667
DWA[85]	MTL	88.405±0.322	−5.452±0.338	44.969±0.378	22.682±0.537	34.290±1.099	−1.053±0.866
PCGrad[86]	MTL	89.698±0.164	−4.921±0.121	47.026±0.538	23.244±0.740	39.427±1.275	2.017±0.769
RLW[83]	MTL	89.288±0.218	−5.226±0.204	47.574±0.349	23.833±0.117	38.531±1.610	1.332±1.148
A-GEM[35]	SCL	87.022±0.519	−7.646±0.483	27.379±0.585	5.416±0.851	28.530±0.994	−7.070±1.410
GMED[72]	SCL	85.471±0.324	−8.875±0.335	49.094±1.792	18.356±1.345	34.495±1.568	−0.640±1.799
ER[87]	SCL	89.106±0.315	−5.525±0.207	47.324±0.584	23.330±0.762	35.950±0.763	−0.767±0.591
MaxDO (AGD)	PCL	90.189±0.314	−4.258±0.311	50.203±0.978	24.510±0.092	40.770±0.354	3.119±0.450

Table 4.2　Class-incremental comparisons (avg±std) on three datasets

Method (+Reherasal)	Type	PS-EMNIST		PS-CIFAR-100		PS-ImageNet-TINY	
		A_e/%	F_e/%	A_e/%	F_e/%	A_e/%	F_e/%
AvgGrad	MTL	43.823±0.566	−44.484±0.596	8.921±0.276	−4.985±0.702	7.511±0.661	−15.825±0.228
MGDA[64]	MTL	52.823±0.201	−7.753±0.538	11.323±0.282	−4.065±0.725	8.318±0.155	−13.118±0.739

continued

Method (+Reherasal)	Type	PS-EMNIST		PS-CIFAR-100		PS-ImageNet-TINY	
		A_e/%	F_e/%	A_e/%	F_e/%	A_e/%	F_e/%
GradNorm[84]	MTL	43.292±0.590	−44.824±0.688	9.477±0.222	−4.403±1.236	6.395±0.130	−15.869±0.336
DWA[85]	MTL	42.734±0.211	−38.944±0.581	3.519±0.116	−7.470±0.244	3.003±0.139	−12.774±0.158
PCGrad[86]	MTL	45.035±0.273	−43.309±0.176	10.140±0.253	−5.149±0.553	7.789±0.237	−15.860±0.318
RLW[83]	MTL	44.595±0.480	−43.424±0.325	9.957±0.135	−5.173±0.613	7.419±0.258	−16.079±0.255
A-GEM[35]	SCL	26.249±0.511	−62.480±0.527	3.374±0.098	−10.462±0.543	3.379±0.149	−18.514±0.516
GMED[72]	SCL	22.694±0.153	−65.805±0.255	4.941±0.407	−12.837±0.710	3.386±0.348	−19.060±0.682
ER[87]	SCL	43.474±0.860	−44.597±0.861	9.317±0.339	−5.337±0.626	6.126±0.277	−13.686±0.290
MaxDO (AGD)	PCL	53.139±0.156	−11.903±0.476	12.237±0.176	−2.280±0.270	9.532±0.363	−12.511±0.610

4.3.4 Rehearsal Analysis in PCL

In Fig. 4.4, we compare the effects of different memory buffer sizes (per class) in rehearsal on PS-CIFAR-100. We observe that the memory buffer size affects the remembering of old knowledge, and a larger size means better knowledge keeping, which is similar to traditional CL. Under the same memory size, our method has better performance. Then, in Table 4.3(a), to show the MaxDO's effectiveness of forgetting reduction on rehearsal gradient, we evaluate the result that only leverages MaxDO on new tasks. In this case, the final gradient is calculated by $d = \frac{1}{2}g_{reh} + \frac{1}{2}g_{new}$, where g_{new} is the solution gradient via MaxDO on only new parallel tasks, and g_{reh} means the gradient from the rehearsal data streams. The result shows that it is necessary to put the rehearsal gradient to the MaxDO. Otherwise, the model will get worse accuracy and forgetting.

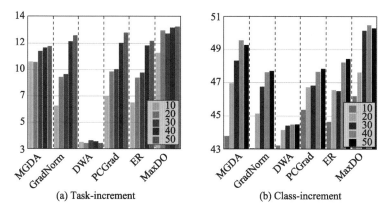

(a) Task-increment (b) Class-increment

Fig. 4.4 Different memory sizes on PS-CIFAR-100

Table 4.3 Rehearsal experiemtns, comparisons with symmetric metrics and ablation studies in MaxDO on PS-CIFAR-100

Experiment	Method	Task-increment		Class-increment	
		$A_e/\%$	$F_e/\%$	$A_e/\%$	$F_e/\%$
(a) Rehearsal analysis	MaxDO (w/o Rehearsal)	25.241± 0.466	3.128± 0.175	2.971± 0.048	−11.356± 0.703
	MaxDO (w/o Reherasal gradient)	47.653± 1.280	23.538± 1.468	9.431± 0.469	−5.907± 0.658
(b) Symmetric metrics	MaxDO (EuDist)	49.201± 0.410	24.112± 0.197	10.189± 0.218	−7.172± 0.796
	MaxDO (CosDist)	49.154± 0.570	25.094± 0.222	11.771± 0.139	−3.114± 0.681
	MaxDO (Norm EuDisto)	48.517± 0.336	24.386± 0.388	11.692± 0.373	−3.228± 0.822
(c) Ablation studies	MaxDO (w/o Max-Discrepancy)	48.957± 0.451	25.088± 0.203	11.323± 0.282	−4.065± 0.725
	MGDA[64]	47.899± 1.022	23.874± 0.447	10.441± 0.324	−4.807± 0.573
	MaxDO (w/o MGDA)	49.631± 0.431	25.816± 0.561	11.881± 0.468	−2.993± 0.964
(d) Our full method	MaxDO (AGD, w/Reherasal)	50.203± 0.274	24.510± 0.092	12.237± 0.176	−2.280+ 0.270

4.3.5 Comparison with Symmetric Metrics

As shown in Table 4.3(b), we compare AGD with three common symmetric metrics including EuDist, CosDist, and Normalized EuDist. EuDist, CosDist are defined in Sec. 4.2.2. The vanilla EuDist depends highly on the gradient magnitude difference, thus we also compare with its normalized version $D(x,y)=\frac{\|x-y\|}{\|x\|+\|y\|}\in[0,1]$, namely Normalized EuDist (Norm EuDist). The results show that the three metrics can also obtain good performance with MaxDO. However, EuDist fails to reduce catastrophic forgetting effectively because of the over-emphasizing of gradient magnitude difference. Considering only the gradient direction difference, MaxDO with CosDist obtains better performance than EuDist. But CosDist ignores the magnitude difference, which is also important in the min-distance problem, resulting in insufficient performance. Compared to EuDist, Norm EuDist obtains even worse performance. In contrast, MaxDO with AGD considers the asymmetric influence on gradient update, and tolerance is set to reduce the influence from small gradients to new-access tasks, which yields the best performance.

4.3.6 Ablation Study

We evaluate the impact of the two main components of MaxDO in Table 4.3(c). First, we block the maximum discrepancy in MaxDO [MaxDO (w/o Max-Discrepancy)], which means that we solve the min-distance problem with Eq. (4.4) directly. Because of the self-interference, the solution combines the minimum discrepancy but fails to effectively reduce the discrepancy from other gradients (48.957% and 11.323% for A_e). We then block the MGDA that obtains a weighted gradient that does not damage any tasks. MGDA is quite useful in traditional MTL tasks but is not suitable in PCL (w/o MGDA, 49.631% and 11.881% for A_e). Because of the diverse training process of parallel tasks, gradients are with large magnitude differences and MGDA prefers to set large factors to small gradients. We solve the problem by both MGDA and the maximum discrepancy, and the whole MaxDO method with AGD outperforms the two ablated methods (50.203% and 12.237% for A_e), where the characters of the two components are combined.

4.3.7 Procedure Time

In Fig. 4.5, we show the training time comparison on PS-CIFAR-100. We first compare the training time for 2 to 5 parallel tasks in one iteration [Fig. 4.5(a)]. We find that the generation of task numbers will grow the training time, and MaxDO needs more time than other methods because multiple minimum distance optimizations are performed. Then, we show both the final accuracy and whole training time in Fig. 4.5(b). In the whole timeline, MaxDO gets slightly longer training time than other methods but better performance.

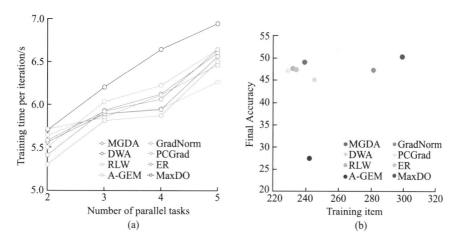

Fig. 4.5 Training time comparisons on PS-CIFAR-100

4.3.8 Tolerance Analysis in AGD

In this chapter, AGD is designed to have a tolerance $\frac{1}{2}$ in Corollary 4.1. This is because updating with a zero gradient will neither improve nor damage the performance. Even though, we prefer positive influence rather than noninfluence. Thus, we define that the distance $\hat{D}(g_A, g_B)$ in the situation $\|g_A\| \ll \|g_B\|$ is the mid-level in the value range. Therefore, we study to change the tolerance and observe the performance change. The tolerance can be controlled by adding a factor $\gamma > 0$ at the denominator. Omitting the edge situation, we have

$$\hat{D}_\gamma(g_A, g_B) = \frac{\|g_A - g_B\|}{\gamma \|g_B\| + \|g_A - g_B\|}.$$

The experiments on different tolerances are shown in Table 4.4. The results show either larger or smaller tolerances compared to $\frac{1}{2}$ will get the performance drop during the PCL training.

Table 4.4 Comparisons on different tolerances (Tol.)

γ	Tol.	Task-increment		Class-increment	
		A_e/%	F_e/%	A_e/%	F_e/%
0.2	5/6	46.118±0.463	22.617±0.484	11.283±0.696	−3.158±1.087
0.5	2/3	46.452±0.113	22.874±0.680	11.284±0.343	−3.305±0.629
1	1/2	50.203±0.274	24.510±0.092	12.237±0.176	−2.280±0.270
2	1/3	49.827±0.420	25.520±0.611	11.603±0.506	−3.503±0.687
3	1/4	48.766±0.171	24.491±0.406	11.389±0.738	−3.234±1.060
4	1/5	48.859±0.627	24.714±0.437	11.378±0.484	−3.189±0.902

4.4 Chapter Conclusion

In this chapter, we studied to address the training conflict and catastrophic forgetting issues in Parallel Continual Learning (PCL). We presented that the two issues are rooted in the gradient discrepancies and formulated the problem into a minimum distance optimization among gradients. However, the distance metric is often set to be symmetric, which is problematic in gradient descent. To evaluate the gradient discrepancy in PCL, we proposed an explicit Asymmetric Gradient Distance (AGD), which considers both gradient magnitude ratios and directions and has a tolerance when updating with a small gradient of inverse direction. Moreover, we proposed a novel Maximum Discrepancy Optimization (MaxDO) strategy to minimize the maximum discrepancy among multiple gradients and avoid self-interference. Solving by MaxDO with AGD, the parallel training in PCL reduces the influence of the training conflict and slows the catastrophic forgetting. We verified the proposed method on three image recognition datasets. The experimental results significantly showed the advantage of MaxDO and the effectiveness of the proposed AGD. We list the latent limitation of our method: (1) The MaxDO cannot guarantee a theoretical Pareto optimum in the training process like MGDA, which means a better trade-off can be obtained in the future. (2) The MaxDO method needs more time for training.

Chapter 5

Multi-Label Continual Learning Using Augmented Graph Convolutional Network

Multi-Label Continual Learning (MLCL) builds a class-incremental framework in a sequential multi-label image recognition data stream. The critical challenges of MLCL are the construction of label relationships on past-missing and future-missing partial labels of training data and the catastrophic forgetting on old classes, resulting in poor generalization. To solve the problems, the study proposes an Augmented Graph Convolutional Network (AGCN) that can construct the cross-task label relationships in MLCL and sustain catastrophic forgetting. First, we build an Augmented Correlation Matrix (ACM) across all seen classes, where the intra-task relationships derive from the hard label statistics. In contrast, the inter-task relationships leverage hard and soft labels from data and a constructed expert network. Then, we propose a novel Partial Label Encoder (PLE) for MLCL, which can extract dynamic class representation for each partial label image as graph nodes and help generate soft labels to create a more convincing ACM and suppress forgetting. Last, to suppress the forgetting of label dependencies across old tasks, we propose a relationship-preserving constrainter to construct label relationships. The inter-class topology can be augmented automatically, which also yields effective class representations. The proposed method is evaluated using two multi-label image benchmarks. The experimental results show that the proposed way is effective for MLCL image recognition and can build convincing correlations across tasks even if the labels of previous tasks are missing.

5.1 Introduction

Machine learning approaches have been reported to exhibit human-level performance on some tasks, such as Atari games[92] or object recognition[93]. However, they always assume that no novel knowledge will be input into models, which is impractical in the real world. To meet the scenario, continual learning develops intelligent systems that can continuously learn new tasks from sequential datasets while preserving learned knowledge of old tasks[94]. Recently, class-incremental continual learning[95] builds an adaptively evolvable classifier for the seen classes at any time, where the learner has no access to the task-ID at inference time[14] just like the real-life applications. Compared to traditional continual learning, a class-incremental model has to distinguish between all seen classes from all tasks. Therefore it is more challenging. For privacy and storage reasons, the training data for old tasks is unavailable when new tasks arrive. As the model incrementally learns new knowledge, old knowledge is overwritten and gets a drop in performance, known as catastrophic forgetting[18]. Thus, the major challenge of MLCL is to learn new tasks without catastrophically forgetting previous tasks over time.

Due to many researchers' efforts, many methods for class-incremental continual learning have been proposed. The rehearsal-based methods[16,35,50,53,70] store samples from raw datasets or generates pseudo-samples with a generative model[36,96], these samples are replayed while learning a new task to prevent forgetting. The regularization-based methods[18,23,35] have an additional regularization term introduced in the loss function, consolidating previous knowledge when learning new tasks. And the parameter isolation methods[45,97] dedicate different model parameters to each task to alleviate any possible forgetting. Moreover, some recent transformer-based methods[33,97,98,98] have also achieved good performance.

However, most existing methods for class-incremental continual learning only consider the input images are single-labelled, and we call them Single-Label Continual Learning (SLCL). SLCL is limited in practical applications such as movie categorization and scene classification, where multi-label data is widely used. As shown in Fig. 5.1, an image contains multiple labels, including "sky", "grass", "person" and "dog", etc., which shows the multi-label space is much larger than the single-label space via label combination. Because of the co-occurrence of multiple labels, the label space of the multi-label dataset is much larger than that of the single-label one. In the recent, using a neural network to tackle Multi-Label (ML) image classification problems has achieved impressive results[74,99], which consider constructing label relationships to improve the classification by using recurrent neural network[100,101] and graph convolutional network[102−104].

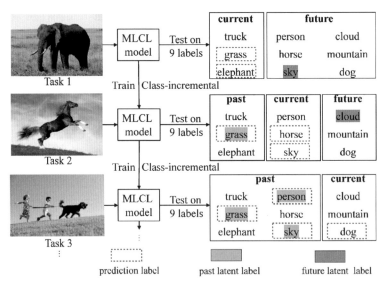

Fig. 5.1 The inference process of MLCL

This chapter puts the multi-task classification into an incremental scenario, i.e. class-incremental classification, and studies how to sequentially learn new classes for Multi-Label Continual Learning (MLCL). The inference process of MLCL is in Fig. 5.1. Given testing images, the model can incrementally recognize multiple labels as new classes are learned continuously. Because of the unavailability of data with future and past classes in continual learning, the partial label problem poses a significant challenge in building multi-label relationships and reducing catastrophic forgetting in MLCL than SLCL. The partial label problem means that each task in MLCL cannot be trained independently since the label spaces for different tasks are overlapped. For example, as shown in Fig. 5.1, the label "sky" is present in all three tasks. It is one of the overlapped labels for three tasks. For Task 1, "sky" is the future latent label, and for Task 3, "sky" is the past latent label. If the past latent label is not annotated in the current training, the past-missing partial label problem will occur, and similarly, the future-missing partial label problem will occur. An MLCL model should incrementally recognize multiple labels as new classes are learned continuously.

Practically, we solve the MLCL problems in two real-world labelling scenarios, i.e. the current training dataset has past and current labels (Continuous Labelling MLCL, CL-MLCL) or only current labels (Independent Labelling MLCL, IL-MLCL). The IL-MLCL has past-missing and future-missing partial label problems, while CL-MLCL has only future-missing partial label problem. As for both scenarios, the partial label problem poses a significant challenge in building multi-label relationships and keeping them from catastrophic forgetting. It is crucial to study a feasible solution to solve the partial label problem in MLCL. This motivates us to design a unified MLCL solution to the sequential multi-label classification problem by considering the label relationships across tasks in both IL-MLCL and CL-MLCL scenarios.

This chapter is an extension of our previous work, Augmented Graph Convolutional Network (AGCN)[105], and we complete the real-world scenario of MLCL (IL-MLCL and CL-MLCL) and propose an improved version, AGCN++. Our AGCN++ has three major parts. First, to relate partial labels across tasks, we propose to construct an Augmented Correlation Matrix (ACM) sequentially in MLCL. We design a unified ACM constructor. For CL-MLCL, ACM is updated by the hard label statistics from new training data at each task. For IL-MLCL, an auto-updated expert network is designed to generate predictions of the old tasks. These predictions are used as soft labels to represent the old classes in constructing ACM. Second, due to partial label problems, effective class representation is difficult to build. In our early conference work, the AGCN model utilized pre-given semantic information (i.e. word embedding) as class representation. The fixed class representation will lead to the accumulation of errors in constructing label relationships due to partial label problems. Then, this will skew predictions and lead to more serious forgetting. So in this work, AGCN++ utilizes a partial label encoder (PLE) to decompose each partial image's feature into dynamic class representations. These class-specific representations will vary from image to image and are input as graph nodes into AGCN++. Moreover, unlike AGCN, which directly adds graph nodes manually, PLE can automatically generate graph nodes for each partial label image. Utilizing PLE to get the graph nodes can also reduce the impact of the low quality of word embeddings. So AGCN++ can generate a more convincing ACM and suppress forgetting. Third, we propose to encode the dynamically constructed ACM and graph nodes. The AGCN++ model correlates the label spaces of both the old and new tasks in a convolutional architecture and mines the latent correlation for every two classes. This information will be combined with the visual features for prediction. Moreover, to further mitigate the forgetting, a distillation loss function and a relationship-preserving graph loss function are designed for class-level forgetting and relationship-level forgetting, respectively.

In this chapter, we construct two multi-label image classification datasets, Split-COCO and Split-WIDE, based on widely-used multi-label datasets MSCOCO and NUS-WIDE. The results on Split-COCO and Split-WIDE show that the proposed AGCN++ effectively reduces catastrophic forgetting for MLCL image recognition and can build convincing correlation across tasks whenever the labels of previous tasks are missing (IL-MLCL) or not (CL-MLCL). Moreover, our methods can effectively reduce catastrophic forgetting in two scenarios.

This chapter extends our AGCN[105] with the following new contents:

We complete the real-world scenario of MLCL from IL-MLCL to CL-MLCL scenarios, and a unified AGCN++ model is redesigned to capture label dependencies to improve multi-label recognition in the data stream.

(1) We propose a novel Partial Label Encoder (PLE) to decompose the global image features into dynamic graph nodes for each partial label image, which reduces the accumulation of errors in the construction of label relationships and suppresses forgetting.

(2) We propose a unified ACM constructor. The ACM is dynamically constructed using soft or hard labels to build label relationships across sequential tasks of MLCL to solve the partial label problem for IL-MLCL and CL-MLCL. The distillation loss and relationship-preserving loss readjust to IL-MLCL and CL-MLCL to mitigate the class- and relationship-level catastrophic forgetting.

(3) More experimental results are provided, including extensive comparisons on two different scenarios settings and more ablation studies, etc. More ablation studies and new SOTA MLCL results are provided.

5.2 Methodology

5.2.1 Definition of MLCL

Given T tasks with respect to training datasets $\{\mathcal{D}_{trn}^1, \cdots, \mathcal{D}_{trn}^T\}$ and test datasets $\{\mathcal{D}_{tst}^1, \cdots, \mathcal{D}_{tst}^T\}$, the total class numbers increase gradually with the sequential tasks in MLCL and the model is constantly learning new knowledge. A continual learning system trains on the training sets from $\mathcal{D}_{trn}^1$ to $\mathcal{D}_{trn}^T$ sequentially and evaluate on all seen test sets at any time. For the t-th new task, the new and task-specific classes are to be trained, namely $\mathcal{C}^t$. MLCL aims at learning a multi-label classifier to discriminate the increasing number of classes in the continual learning process. We denote $\mathcal{C}_{seen}^t = \bigcup_{n=1}^{t} \mathcal{C}^n$ as seen classes at Task t, which contains old class set $\mathcal{C}_{seen}^{t-1}$ and new class set $\mathcal{C}^t$, that is, $\mathcal{C}_{seen}^t = \mathcal{C}_{seen}^{t-1} \cup \mathcal{C}^t$, and $\mathcal{C}_{seen}^{t-1} \cap \mathcal{C}^t = \varnothing$.

5.2.2 MLCL Scenarios

In this section, considering academic and practical requirements, we introduce the two scenarios in MLCL. In one scenario, we adopt strict continual learning and cannot obtain the old class like most single-label continual learning methods[36,80,96,106]. In the other scenario, we consider the real-world setting. IL-MLCL setting has hard task boundaries, so the old classes are unavailable. Conversely, similar to the settings in Bang, Kim, Yoo, et al.[17] and Aljundi, Lin, Goujaud, et al.[90], CL-MLCL setup makes the task boundaries faint. It is closer to the real world, where new classes do not show up exclusively. The difference between the two scenarios is the training label space for old classes.

5.2.2.1 Continuous labelling (CL-MLCL)

CL-MLCL is a more realistic scenario where the data distribution shifts gradually without hard task boundaries. The annotator of CL-MLCL needs to label all seen classes. The class numbers of training data increase gradually with the sequential tasks, i.e. $\mathcal{C}_{seen}^t$ for training data of Task t. As shown in Table 5.1, the label space $y \subseteq C_{seen}^t$. The past latent label is annotated, so the old and new labels coexist for a current sample in CL-MLCL. This scenario is labor-costly, especially when the class number is large. Because the past latent label is annotated, only the future-missing partial label problem will occur in CL-MLCL, and no past-missing partial label problem will occur.

Table 5.1 Training and testing label sets of Task t
in two scenarios, CL-MLCL and IL-MLCL

	CL-MLCL	IL-MLCL
Trian	$\mathscr{C}_{seen} = \mathscr{C}_{seen}^{-1} \cup \mathscr{C}$	$\mathscr{C}$
Test	$\mathscr{C}_{seen} = \mathscr{C}_{seen}^{-1} \cup \mathscr{C}$	$\mathscr{C}_{seen} = \mathscr{C}_{seen}^{-1} \cup \mathscr{C}$

5.2.2.2 Independent labelling (IL-MLCL)

In this scenario, the annotator only labels the new classes in $\mathscr{C}$ for training data in Task t, as shown in Table 5.1. This means the training label space is independently labelled with sequential class-incremental tasks. The old and new labels do not overlap in new task samples in IL-MLCL. The training label space $\mathscr{Y}$ of IL-MLCL at Task t is right the task-specific label (new labels) set $\mathscr{C}$. IL-MLCL can reduce the labelling cost, but due to the lack of old labels in the IL-MLCL label space, the past latent label is not annotated. As a result, a past-missing partial label problem will be caused together with future-missing partial label.

5.2.2.3 Test phase and the goal

During the test phase, the ground truth for each data point contains all the old classes $\mathscr{C}_{seen}^{-1}$ and task-specific classes $\mathscr{C}$ for both CL-MLCL and IL-MLCL. That is, as shown in Table 5.1, the label space in the test phase is the all seen classes $\mathscr{C}_{seen}$. This chapter aims to propose a unified approach to solve the MLCL problem in both IL-MLCL and CL-MLCL scenarios.

5.2.3 Overview of the Proposed Method

In multi-label learning, label relationships are verified effective to improve the recognition[102-104]. However, it is challenging to construct convincing label relationships in MLCL image recognition because of the partial label problem. The partial label problem results in difficulty in constructing the inter-task label relationships. Moreover, forgetting happens not only at the class level but also at the relationship level, which may damage performance.

For effective multi-label recognition, we propose an AGCN++ to construct and update the intra- and inter-task label relationships during the training process (Fig. 5.2). As shown in Fig. 5.2(a), AGCN++ model is mainly composed of three parts: (1) Partial Label Encoder (PLE) decomposes the image feature extracted by the CNN into a group of class-specific representations, these representations are used as graph nodes to feed the GCN model. (2) Augmented Correlation Matrix (ACM) provides the label relationships among all seen classes $\mathscr{C}_{seen}^{t}$ and is augmented to capture the intra- and inter-task label dependencies. (3) Graph Convolutional Network encodes ACM and graph nodes H^t into label representations $\hat{y}_{gph}$ for label relationships. We construct auto-updated expert networks consisting of CNN_{xpt} and GCN_{xpt}. After each task has been trained, the model is saved as the expert model to provide soft labels $\hat{z}$.

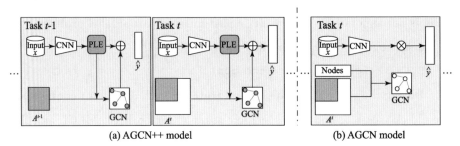

(a) AGCN++ model

(b) AGCN model

(a) AGCN++ model. The model is mainly composed of three points: PLE, ACM, and stacked GCN encoder. PLE decomposes the image feature extracted by the CNN into a group of class-specific representations. The stacked GCN encodes these representations and ACM into graph features. $\hat{y}$ denotes the class-incremental prediction scores. (b) Compared with AGCN++, AGCN directly uses word embedding as graph nodes. Graph nodes and ACM are input into GCN. The graph feature and image feature do the matrix multiplication to get the prediction.

Fig. 5.2 The overall framework of AGCN++ and AGCN

As shown in Fig. 5.2(a) and (b), the most significant difference between AGCN and AGCN++ is that AGCN++ can extract graph nodes from the original image through PLE. GCN encodes ACM and graph nodes to get $\hat{y}_{gph}$. By adding $\hat{y}_{cal}$ and $\hat{y}_{gph}$, the soft label generated by the model can better replace the past-missing partial label, more convincing ACM

(Fig. 5.8) can be developed for IL-MLCL, and the forgetting of CL-MLCL and IL-MLCL can be reduced through knowledge distillation. These can improve the performance of the model.

5.2.4 Partial Label Encoder

Due to the partial label problems in MLCL, effective class representation is difficult to build. In AGCN, the model utilized pre-given word embedding as fixed class representation, which will lead to the accumulation of errors in the construction of label relationships. Then, this will skew predictions and lead to more serious forgetting.

Inspired by Zhou, Khosla, Lapedriza, et al.[107], we propose the Partial Label Encoder (PLE), which decomposes the global image features for each partial label into dynamic representations continuously as classes increment. These class-specific representations will vary from image and are used as augmented graph nodes. PLE will reduce the accumulation of errors in the construction of label relationships and suppress forgetting. And the resulting graph nodes are automatically augmented as the number of classes increases in MLCL.

PLE initializes with image features and model parameters and continuously updates graph nodes. As shown in Fig 5.3, $CNN(x) \in \mathbb{R}^D$, $CNN_{xpt}(x) \in \mathbb{R}^D$, D represents the image feature dimensionality.

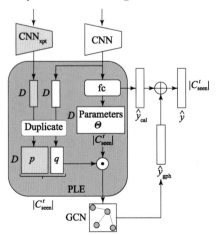

Fig. 5.3　Partial Label Encoder (PLE) in Task t

We use a fully connected layer $fc(\cdot)$ to achieve two goals. One is to get the prediction without adding label dependencies $\hat{y}_{cal}$

$$\hat{y}_{cal} = fc(CNN(x)) \in \mathbb{R}^{|\mathcal{C}_{seen}|} \tag{5.1}$$

The other is to make the image feature aware of class information by doing Hadamard Product with its parameters.

$$H^t = \Theta \odot \text{cat}(p, q) \in \mathbb{R}^{|\mathcal{C}_{seen}^t| \times D}, \tag{5.2}$$

where $\odot$ is the Hadamard Product. p and q are multiple copies of respective image features. $p = \text{Duplicate}(CNN_{xpt}(x)) \in \mathbb{R}^{|\mathcal{C}_{seen}^{t-1}| \times D}$, $q = \text{Duplicate}(CNN(x)) \in \mathbb{R}^{|a| \times D}$. For example, $CNN(x) \in \mathbb{R}^{1 \times D}$ is copied D times to get $q \in \mathbb{R}^{|a| \times D}$. $\Theta \in \mathbb{R}^{|\mathcal{C}_{seen}| \times D}$ represents the class-specific fully-connected layer parameters. The dimension of Θ is continuously expanded to accommodate the class-incremental characteristic in continual learning. In Eq. (5.2), H^t represents the class-aware graph node and automatically augments as the new task progresses. We then encode H^t by Graph Convolutional Network (GCN) to get graph representation $\hat{y}_{gph}$.

$$\hat{y}_{gph} = GCN(A^t, H^t) \in \mathbb{R}^{|\mathcal{C}_{seen}|}, \tag{5.3}$$

where A^t denotes the Augmented Correlation Matrix (ACM). GCN is a two-layer stacked graph model similar to MLGCN[102,104]. ACM A^t and graph node H^t can be augmented as the class number increments. With the established ACM, GCN provides dynamic label relationships to CNN for prediction.

Moreover, we introduce the prediction $\hat{y}_{cal}$ without adding label dependencies, which is combined with $\hat{y}_{gph}$ as the final multi-label prediction $\hat{y} \in \mathbb{R}^{|\mathcal{C}_{seen}|}$ of our model:

$$\hat{y} = \sigma(\hat{y}_{cal} + \hat{y}_{gph}), \tag{5.4}$$

where $\sigma(\cdot)$ represents the Sigmoid function.

ACM represents the auto-updated dependency among all seen classes in the MLCL image recognition system. The next section will introduce how to establish and augment ACM in AGCN++.

5.2.5 Augmented Correlation Matrix

Most existing multi-label learning algorithms[102-104] rely on constructing the inferring label correlation matrix $\boldsymbol{A}$ by the hard label statistics among the class set $\mathscr{C}$: $\boldsymbol{A}_{ij} = P(\mathscr{C}_i | \mathscr{C}_j)|_{i \neq j}$. The correlation matrix represents a fully-connected graph. When a new task comes, the graph should be augmented automatically. However, in MLCL, the label correlation matrix is hard to infer directly by statistics because of the partial label problem.

To tackle the problem, as shown in Fig. 5.4(b), we introduce an autoupdated expert network inspired by Li, Song and Luo[74] and Zhou, Ye and Zhan[97], which is used to provide missing past labels. The soft labels $\hat{z}$ is obtained by feeding data of the current task into the expert model, i.e. $\hat{z} = \text{expert}(x)$. Based on the soft labels, as shown in Fig. 5.4 (a), we construct an Augmented Correlation Matrix (ACM) $\boldsymbol{A}^t$ in IL-MLCL and CL-MLCL:

$$\boldsymbol{A}^t = \begin{bmatrix} \boldsymbol{A}^{t-1} & \boldsymbol{R}^t \\ \boldsymbol{Q}^t & \boldsymbol{B}^t \end{bmatrix} \Leftrightarrow \begin{bmatrix} \text{Old-Old} & \text{Old-New} \\ \text{New-Old} & \text{New-New} \end{bmatrix}, \quad (5.5)$$

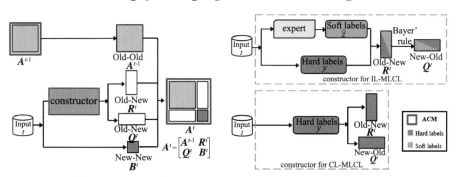

(a) The unified construction of ACM (b) The ACM constructor for IL-MLCL and CL-MLCL

(a) The unified construction of ACM $\boldsymbol{A}^t$ using Eq. (5.5). $\boldsymbol{A}^{t-1}$ is the completely-built ACM of Task $t-1$ in two scenarios. (b) The ACM constructor for IL-MLCL and CL-MLCL. $\boldsymbol{B}^t$ is constructed using hard labels from a dataset of Task t using Eq. (5.6); the $\boldsymbol{R}^t$ is constructed by hard labels from a dataset of Task t and soft labels from the expert using Eq. (5.7), the expert is the saved model of Task $t-1$; the $\boldsymbol{Q}^t$ is constructed by $\boldsymbol{R}^t$ based on Bayes' rule using Eq. (5.8). For CL-MLCL, the ACM is constructed using hard label statistics in four blocks, including Old-Old, New-New, Old-New and New-Old blocks by Eq. (5.6), Eq. (5.7) and Eq. (5.8).

Fig. 5.4 The construction of ACM

in which we take four block matrices including A^{t-1} and B^t, R^t and Q^t to represent intra- and inter-task label relationships between old and old classes, new and new classes, old and new classes as well as new and old classes respectively. For the first task, $A^1 = B^1$. For $t > 1$, $A^t \in \mathbb{R}^{|\mathscr{C}_{seen}| \times |\mathscr{C}_{seen}|}$. It is worth noting that the block A^{t-1} (Old-Old) can be derived from the old task, so we focus on how to compute the other three blocks in the ACM.

New-New block ($B^t \in \mathbb{R}^{|\mathscr{C}^a| \times |\mathscr{C}^a|}$). As shown in Fig. 5.4 (a), this block computes the intra-task label relationships among the new classes, and the conditional probability in B^t can be calculated using the hard label statistics from the training dataset similar to the common multi-label learning:

$$B^t_{ij} = P(\mathscr{C}_i \in \mathscr{C} \mid \mathscr{C}_j \in \mathscr{C}) = \frac{N_{ij}}{N_j}, \quad (5.6)$$

where N_{ij} is the number of examples with both class $\mathscr{C}_i$ and $\mathscr{C}_j$, N_j is the number of examples with class $\mathscr{C}_j^a$. Due to the data stream, N_{ij} and N_j are accumulated and updated at each step of the training process. This block is shared by both IL-MLCL and CL-MLCL because the new class data is always available.

Old-New block ($R^t \in \mathbb{R}^{|\mathscr{C}_{seen}^{t-1}| \times |\mathscr{C}^a|}$). As shown in Fig. 5.4 (b), for CL-MLCL, this block can be directly obtained by the hard label statistics. For IL-MLCL, given an image x, for old classes, $\hat{z}_i$ (predicted probability) generated by the expert can be considered as the soft label for the i-th class. Thus, the product $\hat{z}_i y_j$ can be regarded as an alternative of the cooccurrences of $\mathscr{C}_{seeni}^{t-1}$ and $\mathscr{C}_j$. Thus, $\sum_x \hat{z}_i y_j$ means the online mini-batch accumulation

$$R^t_{ij} = P(\mathscr{C}_{seeni}^{t-1} \in \mathscr{C}_{seen}^{t-1} \mid \mathscr{C}_j \in \mathscr{C}) = \begin{cases} \dfrac{N_{ij}}{N_j}, & \text{if CL-MLCL,} \\ \dfrac{\sum_x \hat{z}_i y_j}{N_j}, & \text{if IL-MLCL,} \end{cases} \quad (5.7)$$

where N_{ij} is the accumulated number of examples with both class $\mathscr{C}_{seeni}^{t-1}$ and

$\mathscr{C}_j$, and N_j is the accumulated number of examples with class $\mathscr{C}_j$.

New-Old block ($\boldsymbol{Q}^t \in \mathbb{R}^{|\mathscr{C}| \times |\mathscr{C}_{\text{seen}}^{-1}|}$). As shown in Fig. 5.4 (b), for CL-MLCL, the inter-task relationship between new and old classes can be computed using hard label statistics. For IL-MLCL, based on the Bayes' rule, we can obtain this block by

$$\boldsymbol{Q}_{ji}^t = P(\mathscr{C}_j \in \mathscr{C} \mid \mathscr{C}_{\text{seen}i}^{-1} \in \mathscr{C}_{\text{seen}}^{-1})$$

$$= \begin{cases} \dfrac{N_{ij}}{N_i}, & \text{if CL-MLCL}, \\ \dfrac{P(\mathscr{C}_{\text{seen}i}^{-1} \mid \mathscr{C}_j) P(\mathscr{C}_j)}{P(\mathscr{C}_{\text{seen}i}^{-1})} = \dfrac{\boldsymbol{R}_{ij}^t N_j}{\sum_x \hat{z}_i}, & \text{if IL-MLCL}, \end{cases} \quad (5.8)$$

where N_{ij} is the accumulated number of examples with both class $\mathscr{C}_{\text{seen}i}^{-1}$ and $\mathscr{C}_j$, N_i is the accumulated number of examples with class $\mathscr{C}_{\text{seen}i}^{-1}$.

Finally, we construct an ACM using the soft and hard label statistics (IL-MLCL) or only the hard label statistics (CL-MLCL). Based on the established ACM, the GCN can capture the label dependencies across different tasks, improving the performance of continual multi-label recognition tasks.

5.2.6 Objective Function

As mentioned above, the class-incremental prediction scores $\hat{y}$ for an image x can be calculated by Eq. (5.4). The prediction $\hat{y} = [\hat{y}_{\text{old}}, \hat{y}_{\text{new}}] \in \mathbb{R}^{|\mathscr{C}_{\text{seen}}|}$, where $\hat{y}_{\text{old}} \in \mathbb{R}^{|\mathscr{C}_{\text{seen}}^{-1}|}$ for old classes and $\hat{y}_{\text{new}} \in \mathbb{R}^{|\mathscr{C}|}$ for new classes when $\mathscr{C}_{\text{seen}} = \mathscr{C}_{\text{seen}}^{-1} \cup \mathscr{C}$. By binarizing the ground truth to hard labels $y = [y_1, \cdots, y_{|\mathscr{C}|}]$, $y_i \in \{0, 1\}$, we train the current task using the Cross Entropy loss

$$\ell_{\text{cls}}(y, \hat{y}) = -\sum_{i=1}^{|\mathscr{C}|} [y_i \log(\hat{y}_i) + (1 - y_i) \log(1 - \hat{y}_i)], \quad (5.9)$$

where $\mathscr{C} = \mathscr{C}_{\text{seen}}$ in CL-MLCL and $\mathscr{C} = \mathscr{C}$ in IL-MLCL. However, like traditional SLCL, sequentially fine-tuning the model on the current task will lead to class-level forgetting of the old classes. To mitigate the class-level catastrophic forgetting, based on the expert network, we construct the distillation loss as

$$l_{dst}(\hat{z}, \hat{y}_{old}) = - \sum_{i=1}^{|\mathcal{C}_{seen}^{t-1}|} [\hat{z}_i \log(\hat{y}_i) + (1-\hat{z}_i)\log(1-\hat{y}_i)], \quad (5.10)$$

where $\hat{z}$ is the soft labels used to represent the prediction on old classes. The soft labels $\hat{z}$ are used to be the target feature for the old class prediction $\hat{y}_{old}$. Soft labels play two main roles in our chapter: (1) $\hat{z}$ are used to be the target feature of old classes to mitigate the class-level forgetting in IL-MLCL and CL-MLCL scenarios; (2) IL-MLCL has the past-missing partial label problem, soft labels are used as substitutes for old class labels to build label relationships across new and old classes.

Algorithm 4: Training procedure of AGCN++

Input: $\mathcal{D}_{trn}^t$

1 for $t = 1: T$ do
2   for $(x, y) \sim \mathcal{D}_{trn}^t$ do
3   if $t = 1$ then
4   Compute $\mathbf{A}^1$ with y using Eq.(5.6);
5              $\mathbf{H}^1, \hat{y}_{cal} = \text{PLE}(\text{CNN}(x))$;
6              $\hat{y}_{gph} = \text{GCN}(\mathbf{A}^1, \mathbf{H}^1)$
7              $\hat{y} = \sigma(\hat{y}_{cal} \oplus \hat{y}_{gph})$
8              $l = l_{cls}(y, \hat{y})$
9   else
10             $\hat{z} = \text{expert}(x)$;
                // get soft labels.
11   Compute $\mathbf{B}^t$ with y using Eq.(5.6);
12   Compute $\mathbf{R}^t$ and $\mathbf{Q}^t$ using Eq.(5.7) and (5.8);
13             $\mathbf{A}^t = \begin{bmatrix} \mathbf{A}^{t-1} & \mathbf{R}^t \\ \mathbf{Q}^t & \mathbf{B}^t \end{bmatrix}$;
                // compute ACM of Task t.
14             $\mathbf{H}^t, \hat{y}_{cal} = \text{PLE}(\text{CNN}(x))$;
15             $\hat{y}_{gph} = \text{GCN}(\mathbf{A}^t, \mathbf{H}^t)$;
                // get new graph represnetation.
16             $y'_{gph} = \text{GCN}_{xpt}(\mathbf{A}^{t-1}, \mathbf{H}^{t-1})$;
                // get target represnetation.
17             $\hat{y} = \sigma(\hat{y}_{cal} \oplus \hat{y}_{gph})$;

18		//get class prediction.
		 $l = \lambda_1 l_{cls}(y,\hat{y}) + \lambda_2 l_{dst}(\hat{z},\hat{y}_{old}) + \lambda_3 l_{gph}(y'_{gph},\hat{y}_{gph})$;
		 //compute the final loss.

9		end
21		Update AGCN++ model by minimizing l
22	end	
23		Update expert model //save parameters to the expert model.
24	end	

To mitigate relationship-level forgetting across tasks, we constantly preserve the established relationships in the sequential tasks. We compute the old graph representation to serve as a teacher to guide the training of the new GCN model. The old graph representation in Task t is computed by $y'_{gph} = \text{GCN}_{xpt}(A^{t-1}, H^{t-1}), t > 1$. Then, we propose a relationship-preserving loss as a relationship constraint:

$$l_{gph}(y'_{gph},\hat{y}_{gph}) = \sum_{i=1}^{|\mathscr{C}_{seen}^{t-1}|} \| y'_{gph,i} - \hat{y}_{gph,i} \|^2. \quad (5.11)$$

By minimizing l_{gph} with the partial constraint of old node embedding, the changes of GCN parameters are limited. Thus, the forgetting of the established label relationships is alleviated with the progress of MLCL image recognition.

The final loss for the IL-MLCL and CL-MLCL model training is defined as

$$l = \lambda_1 l_{cls}(y,\hat{y}) + \lambda_2 l_{dst}(\hat{z},\hat{y}_{old}) + \lambda_3 l_{gph}(y'_{gph},\hat{y}_{gph}), \quad (5.12)$$

where l_{cls} is the classification loss, l_{dst} is used to mitigate the class-level forgetting and l_{gph} is used to reduce the relationship-level forgetting. λ_1, λ_2 and λ_3 are the loss weights for l_{cls}, l_{dst} and l_{gph}, respectively.

The AGCN++ algorithm for both IL-MLCL and CL-MLCL scenarios is presented in Algorithm 4 to show the detailed training procedure. Given the training dataset $\mathscr{D}_{trn}^t$: (1) For the first task, the intra-task correlation matrix A^1 is constructed by the statistics of hard labels y. After the input x is fed to the CNN, the class-specific feature $\hat{y}_{cal}$ and the graph nodes H^1 is obtained by PLE. Then the GCN encodes $\boldsymbol{A}^1$ and $\boldsymbol{H}^1$ to get graph representation $\hat{y}_{gph}$. The prediction score $\hat{y}$ is generated by $\hat{y}_{cal}$ and $\hat{y}_{gph}$. (2) When $t>1$, the ACM $\boldsymbol{A}^t$ is augmented via soft labels $\hat{z}$ and the Bayes' rule. Based on the $\boldsymbol{A}^t$, GCN model can capture both intra- and inter-task label dependencies. Then, $\hat{z}$ and $\hat{y}_{gph}$ as target features to build l_{dst} and l_{gph} respectively. (3) The AGCN++ and expert models are updated respectively.

5.3 Experiments

5.3.1 Datasets

5.3.1.1 Dataset description

We use two datasets, Split-COCO and Split-WIDE, to evaluate the effectiveness of the proposed method. (1) Split-COCO. We choose the 40 most frequent concepts from 80 classes of MS-COCO[110] to construct Split-COCO, which has 65,082 examples for training and 27,173 examples for validation. The 40 classes are split into ten different and non-overlapping tasks, each containing four classes. (2) Split-WIDE. NUS-WIDE[111] is a raw web-crawled multi-label image dataset. We further curate a sequential class-incremental dataset from NUSWIDE. Following Jiang and Li[112], we choose the 21 most frequent concepts from 81 classes of NUS-WIDE to construct the Split-WIDE, which has 144,858 examples for training and 41,146 examples for validation. Split-WIDE has a larger scale than Split-COCO. We split the Split-WIDE into 7 tasks, where each task contains 3 classes.

5.3.1.2 Dataset collection

We enlist the curation details of Split-COCO and Split-WIDE. In the previous continual learning methods, Shmelkov et al.[113] selects 20 out of 80 classes to create 2 tasks for SLCL, each with 10 classes. Nguyen et al.[114] tailor MS-COCO for continual learning of captioning. They select 24 out of 80 classes to create two tasks. PRS[106] needs more low-frequency classes to study the imbalanced problem. They curate four tasks with 70 classes using MS-COCO. Compared to these previous splitting, on the one hand, we set more tasks to test the robustness of the algorithm over more tasks to create a continual setting. On the other hand, we selected more frequent concepts from the original dataset to reduce the long-tail effect of the original data. Multi-label datasets inherently have intersecting concepts among the data points. Hence, a naive splitting strategy may lead to a dangerous amount of data loss. This motivates us to minimize data loss during the split. Moreover, to test diverse research environments, the second objective is to keep the size of the splits balanced optionally. To split the well-known MS-COCO and NUS-WIDE into several different tasks fairly and uniformly, we introduce two kinds of labelling in the datasets. (1) Specific-labelling: if an image only has the labels that belong to the task-special class set $\mathscr{C}^t$ of Task t, we regard it as a specific labelling image for Task t; (2) Mixed-labelling: if an image not only has the task-specific labels but also has the old labels belonging to the class set $\mathscr{C}_{seen}^{t-1}$, we regard it as a mixed-labelling image.

In IL-MLCL, because the model learns from the task-specific labels $\mathscr{C}^t$, the training data is labelled without old labels, so IL-MLCL will suffer from the partial label problem, which mainly appears in the mixed-labelling image. The IL-MLCL and CL-MLCL share the same training images. A randomly data-splitting approach may lead to the imbalance of specific-labelling and mixed-labelling images for each task. We split two datasets into sequential tasks with the following strategies to ensure a proper proportion. We first count the number of labels for each image. Then, we give priority to leaving specific-label images for each task. The mixed-labelling images are then allocated to other tasks. The dataset construction is presented in Fig 5.5.

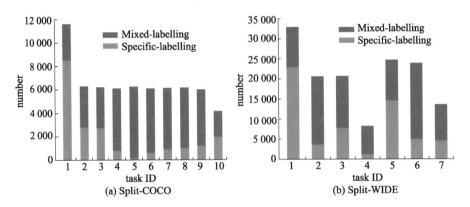

Fig. 5.5 Dataset construction of Split-COCO and Split-WIDE

5.3.2 Evaluation Metrics

5.3.2.1 Multi-label evaluation

Following these multi-label learning methods[102-104], 7 metrics are leveraged in MLCL. (1) the average precision (AP) on each label and the mean average precision (mAP) over all labels; (2) the per-class F1-measure (CF1); (3) the overall F1-measure (OF1). The mAP, CF1 and OF1 are relatively more important for multi-label performance evaluation. Moreover, we adopt 4 other metrics: per-class precision (CP), per-class recall (CR), overall precision (OP) and overall recall (CR).

$$OP = \frac{\sum_i N_i^c}{\sum_i N_i^p}, \quad OR = \frac{\sum_i N_i^c}{\sum_i N_i^g}, \quad OF1 = \frac{2 \times OP \times OR}{OP + OR}$$

$$CP = \frac{1}{C} \sum_i \frac{N_i^c}{N_i^p}, \quad CR = \frac{1}{C} \sum_i \frac{N_i^c}{N_i^g}, \quad CF1 = \frac{2 \times CP \times CR}{CP + CR},$$

where i is the class label and C is the number of labels. N_i^c is the number of correctly predicted images for class i, N_i^p is the number of predicted images for class i and N_i^g is the number of ground-truth for class i.

5.3.2.2 Forgetting measure[59]

This metric denotes the above multi-label metric value difference for each task between testing when it was first trained, and the last task trained. For example, the forgetting measure of mAP for Task t can be computed by its performance difference between Task T and t was trained. F_t, average forgetting after the model has been trained continually up till Task $t \in \{1, \cdots, T\}$ is defined as

$$F_t = \frac{1}{t-1} \sum_{j=1}^{t-1} f_j^t, \qquad (5.13)$$

where f_j^t is the forgetting on Task j after the model is trained up till Task t and computed as

$$f_j^t = \max_{l \in \{1, \cdots, k-1\}} a_{l,j} - a_{t,j}, \qquad (5.14)$$

where a denotes every metric in MLCL like mAP, CF1 and OF1. We evaluate the final forgetting (F_T) after training the final task.

5.3.3 Implementation Details

Following existing multi-label image classification methods[102,103,106], we employ ResNet101[12] as the image feature extractor pre-trained on ImageNet[68]. We adopt Adam[115] as the optimizer of network with $\beta_1 = 0.9, \beta_2 = 0.999$, and $\varepsilon = 10^{-4}$. Following Chen, Wei, Wang, et al.[102], Chen, Lin, Hui, et al.[103], our AGCN++ consists of two GCN layers with output dimensionality of 1024 and 2048, respectively. The input images are randomly cropped and resized to 448×448 with random horizontal flips for data augmentation. The network is trained for a single epoch like most continual learning methods done[17,80,95,116].

5.3.4 Baseline Methods

MLCL is a new paradigm of continual learning. We compare our method with several essential and state-of-art continual learning methods, including (1) EWC[18], which regularizes the training loss to avoid catastrophic forgetting; (2) LwF[74], which uses the distillation loss by saving task-specific parameters; (3) ER[87], which saves a few training data from the old tasks and retrains them in the current training; (4) A-GEM[35] resets the training gradient by combining the gradient on the memory and training data; (5) PRS[106], which uses an improved reservoir sampling strategy to study the imbalanced problem. PRS studies similar problems with us. Still, they focus more on the imbalanced problem but ignore the label relationships and the problem of partial labels for MLCL image recognition. (6) SCR[109], which proposes the NCM classifier to improve SLCL performance. SCR is an algorithm designed to improve the top-1 accuracy of single-label recognition. Similar to Zhou, Ye, and Zhan[97], Kim, Jeong, and Kim[106], Mai, Li, Kim, et al.[109], we use a Multi-Task baseline, which is trained on a single pass over shuffled data from all tasks. It can be seen as the performance upper bound. We also compare with the Fine-Tuning, which performs training without any continual learning technique. Thus, it can be regarded as the performance lower bound. Note that, to extend some SLCL methods to MLCL, we turn the final Softmax layer in each of these methods into a Sigmoid. Other details follow their original settings.

5.3.5 Main Results

5.3.5.1 Split-WIDE results

In Table 5.2, with the establishment of relationships and inhibition of class level and relationship-level forgetting using distillation and relationship preserving loss, our method shows better performance than the other state-of-art performances in both IL-MLCL and CL-MLCL scenarios. In particular, AGCN and AGCN++ perform better than other comparison methods on three more essential evaluation metrics, including mAP, CF1 and OF1, which means the effectiveness in multi-label classification. Also, in the forgetting value evaluated after Task T, we achieve a better forgetting measure, which means the stability of the proposed method in MLCL. In the IL-MLCL scenario, because we use soft labels to replace hard labels in the old task label space and establish and remember the label relationships, the AGCN++ outperforms the most state-of-art performances by a large margin: 45.73% vs. 42.15% (+3.58%) on mAP, 43.04% vs. 37.99% (+5.05%) on CF1 and 45.26% vs. 43.70% (+1.56%) on OF1, as shown in Table 5.2. Like IL-MLCL, CL-MLCL still needs to model complete label dependencies between label relationships and reduce forgetting. The AGCN++ shows better performance than the others in CL-MLCL: 57.07% vs. 54.20% (+2.87%) on mAP, 54.66% vs. 46.13% (+8.53%) on CF1 and 59.29% vs. 55.35% (+3.94%) on OF1, as demonstrated in Table 5.2, which suggests that AGCN++ is effective in a large-scale multi-label dataset.

Table 5.2　The comparisons on the 3-way Split-WIDE dataset

Method	Split-WIDE IL-MLCL							Split-WIDE CL-MLCL						
	mAP↑	CP↑	CR↑	CF1↑	OP↑	OR↑	OF1↑	mAP↑	CP↑	CR↑	CF1↑	OP↑	OR↑	OF1↑
Multi-Task	66.17	69.15	55.30	61.45	77.74	66.30	71.57	69.19	60.60	43.96	50.33	78.45	59.39	66.67
Fine-Tuning	20.33	15.21	37.85	19.10	25.15	61.62	35.72	41.82	44.48	34.73	39.00	52.94	42.97	47.43
Forgetting↓	40.85	36.95	27.20	31.20	27.22	12.14	15.10	18.20	20.23	41.83	28.07	3.29	16.63	11.56
EWC[18]	22.03	15.99	39.53	22.78	24.92	62.97	35.70	45.04	45.33	37.13	40.82	54.36	53.81	54.08
Forgetting↓	34.86	35.51	24.23	28.18	28.41	9.55	15.17	14.73	18.31	38.72	25.78	2.20	7.09	4.06
LwF[74]	29.46	21.65	46.96	29.64	30.77	69.70	42.69	46.44	51.05	33.01	40.09	54.24	46.40	50.01
Forgetting↓	20.26	29.21	17.02	18.99	20.94	3.84	5.73	12.68	9.08	42.93	26.05	2.23	14.33	9.31
A-GEM[35]	32.47	23.26	58.44	33.28	26.36	74.40	38.93	46.83	50.48	27.67	35.75	47.93	35.18	40.58
Forgetting↓	16.42	28.09	8.67	15.71	26.55	6.95	9.73	11.91	10.05	48.36	33.55	11.21	21.56	17.22
ER[108]	34.03	24.64	60.02	34.94	26.62	75.57	39.37	48.08	54.33	31.16	39.61	53.40	38.84	44.98
Forgetting↓	15.15	26.18	7.14	11.80	26.45	6.25	8.61	9.24	7.13	45.32	27.58	2.96	19.28	14.53
PRS[106]	39.70	52.77	18.24	26.48	60.81	14.05	22.19	51.42	58.26	37.64	45.73	55.66	48.90	52.06
Forgetting↓	11.24	4.08	43.22	34.48	2.34	55.73	43.76	7.86	2.21	37.12	16.68	1.90	11.36	7.13
SCR[109]	35.34	28.33	54.34	35.47	32.21	70.28	41.92	49.23	53.87	36.86	43.77	50.16	47.58	48.84
Forgetting↓	14.26	21.29	9.56	10.17	23.09	7.26	8.04	8.34	7.89	39.22	20.56	6.62	13.56	10.78
AGCN	42.15	26.04	70.21	37.99	29.53	84.02	43.70	54.20	56.24	39.10	46.13	53.94	56.84	55.35
Forgetting	10.34	25.44	1.35	5.82	25.23	1.12	4.12	5.27	4.56	34.54	14.34	2.49	5.40	3.26
AGCN++	45.73	33.08	61.57	43.04	32.29	75.62	45.26	57.07	55.07	54.26	54.66	49.03	74.96	59.29
Forgetting	8.32	17.28	7.44	2.13	22.11	5.52	3.98	4.45	3.56	19.48	10.64	6.13	0.18	1.02

Note: We report seven metrics (%) for multi-label classification after the whole data stream is seen once on Split-WIDE in both IL-MLCL and CL-MLCL scenarios. The Multi-Task is offline trained as the upper bound, and Fine-Tuning is the lower bound.

5.3.5.2　Split-COCO results

Split-COCO is split into ten tasks, as mentioned in Lange and

Tuytelaars[116], compared with methods PRS[106] and ER[34], our approach can protect privacy better because AGCN++ does not collect data from the original dataset. As shown in Table 5.3, in IL-MLCL and CL-MLCL, AGCN++ achieves better performance than the others in most metrics. AGCN++ also has a low rate of forgetting old knowledge. With the AGCN++ combining intra- and inter-task label relationships, the proposed AGCN++ outperforms the most state-of-art performances in IL-MLCL: 38.23% vs. 34.11% (+4.12%) on mAP, 41.38% vs. 35.49% (+5.89%) on CF1 and 45.26% vs. 42.37 (+2.89%) on OF1. This means soft labels can effectively replace hard labels in the old task label space to alleviate the partial label problem. AGCN++ is also better in CL-MLCL: 53.49% vs. 48.82% (+4.67%) on mAP, 49.55% vs. 39.18% (+10.37%) on CF1 and 59.32% vs. 56.76% (+2.56%) on OF1. This means ACM is effective for both IL-MLCL and CL-MLCL scenarios in Split-COCO. As illustrated above, AGCN++ can be a uniform MLCL method for IL-MLCL and CL-MLCL.

Table 5.3 The comparisons on the 4-way Split-COCO dataset

Method	Split-COCO IL-MLCL							Split-COCO CL-MLCL						
	mAP↑	CP↑	CR↑	CF1↑	OP↑	OR↑	OF1↑	mAP↑	CP↑	CR↑	CF1↑	OP↑	OR↑	OF1↑
Multi-Task	65.85	71.64	54.31	61.79	77.24	58.03	66.27	68.33	73.06	49.82	61.49	89.31	64.99	74.98
Fine-Tuning	9.83	6.90	18.52	10.54	21.63	41.25	28.83	32.35	33.34	29.10	31.01	57.03	45.56	50.56
Forgetting↓	58.04	48.96	64.30	63.54	18.24	38.76	20.60	31.78	33.32	32.14	33.90	16.61	9.22	12.24
EWC[18]	12.20	9.70	17.54	12.50	23.63	39.84	29.67	35.83	31.88	33.05	32.18	57.62	46.98	51.60
Forgetting↓	45.61	42.68	60.50	55.44	15.34	40.59	19.85	27.66	37.82	26.29	30.29	16.24	7.18	10.16
LwF[74]	19.95	18.02	28.44	21.69	33.14	57.83	40.68	40.87	44.36	35.07	39.07	61.72	48.10	53.95
Forgetting	41.16	29.73	44.01	39.85	8.70	19.38	11.43	21.15	22.70	25.90	23.64	12.29	4.99	7.67
A-GEM[35]	23.31	22.34	42.10	27.25	29.95	62.32	37.94	42.25	64.40	19.28	29.08	57.64	12.62	18.59
Forgetting↓	34.52	17.12	20.36	18.92	13.02	11.35	12.94	19.75	9.11	45.66	35.37	15.94	34.80	39.92
ER[108]	25.03	26.45	41.14	30.54	30.32	61.84	38.38	43.54	71.15	16.65	26.60	62.89	17.72	26.44
Forgetting↓	33.46	14.96	22.28	17.28	11.80	12.49	12.34	17.13	1.14	47.73	38.34	11.79	30.32	32.66
PRS[106]	31.08	56.07	22.74	32.27	57.87	14.24	22.25	46.39	58.56	29.41	38.20	58.84	50.54	54.25
Forgetting↓	28.82	1.32	50.59	16.21	0.34	58.37	30.43	13.07	13.52	31.23	24.56	14.21	6.09	6.36
SCR[109]	25.75	25.22	49.35	30.63	29.40	69.91	39.10	44.96	54.00	29.25	37.82	41.47	40.40	40.46
Forgetting↓	32.02	15.27	16.02	15.98	13.58	6.52	11.96	15.33	19.88	31.89	25.12	30.04	13.24	19.26
AGCN	34.11	31.80	47.73	35.49	34.38	67.72	42.37	48.82	55.73	30.83	39.18	74.27	47.06	56.76

continued

Method	Split-COCO IL-MLCL							Split-COCO CL-MLCL						
	mAP↑	CP↑	CR↑	CF1↑	OP↑	OR↑	OF1↑	mAP↑	CP↑	CR↑	CF1↑	OP↑	OR↑	OF1↑
Forgetting↓	23.71	12.21	17.81	14.79	8.03	9.86	8.16	10.40	18.72	30.39	22.38	1.63	6.71	3.97
AGCN++	38.23	32.51	60.47	41.38	34.32	75.34	45.26	53.49	46.66	52.96	49.55	55.14	64.74	59.32
Forgetting↓	20.12	11.13	9.08	11.34	8.82	3.68	6.78	7.24	24.88	6.34	14.52	23.67	1.34	1.24

Note: We report seven metrics (%) for multi-label classification after the whole data stream is seen once on Split-COCO in both IL-MLCL and CL-MLCL scenarios. The Multi-Task is offline trained as the upper bound, and Fine-Tuning is the lower bound.

5.3.6 More MLCL Settings

In order to prove the robustness of the method, we verify the effectiveness of AGCN and AGCN++ under other MLCL settings. First, as shown in Table 5.4, we increase the number of classes of each task to verify that the proposed PLE and ACM can effectively handle more label relationships in a task: Specifically, 8-way for Split-COCO and 7-way for Split-WIDE. As shown in Table 5.4, our AGCN and AGCN++ can still achieve better results in three more important metrics, mAP, CF1 and OF1. Take the mAP for example. For Split-COCO, 65.92% vs. 62.60% (+3.32%) in IL 8-way and 73.41% vs. 70.24% (+3.17%) in CL 8-way. For Split-WIDE, 56.63% vs. 54.11% (+2.52%) in IL 7-way and 61.04% vs. 58.98% (+2.06%) in CL 7-way. Second, considering in the real world, different tasks often have different numbers of classes. So we provide a different number of classes for each task in a random manner. Specifically, the task setting is "7:4:1:6:2:2:5:7:3:3".

Table 5.4 The comparisons on the 8-way Split-COCO and 7-way Split-WIDE

Method	Split-COCO (IL 8-way)			Split-COCO (CL 8-way)			Split-WIDE (IL 7-way)			Split-WIDE (CL 7-way)		
	mAP	CF1	OF1	mAP	CF1	OF1	mAP	CF1	OF1	mAP	CF1	OF1
Multi-Task	72.24	64.34	72.35	74.63	68.87	79.74	64.71	58.86	53.82	68.84	58.64	60.23
Fine-Tuning	16.89	8.82	46.63	62.69	56.87	68.45	39.26	34.51	50.10	55.46	41.76	57.01
EWC[18]	23.47	11.03	47.82	63.10	56.91	68.60	39.56	33.20	47.69	56.73	51.46	66.07
LwF[74]	36.67	37.38	46.98	63.66	58.39	69.74	40.14	43.64	56.22	57.48	54.90	68.24
A-GEM[35]	40.43	33.50	45.75	63.92	55.30	69.34	42.33	40.66	52.94	57.52	45.36	58.25

Method	Split-COCO (IL 8-way)			Split-COCO (CL 8-way)			Split-WIDE (IL 7-way)			Split-WIDE (CL 7-way)		
	mAP	CF1	OF1	mAP	CF1	OF1	mAP	CF1	OF1	mAP	CF1	OF1
ER[87]	42.23	39.89	46.24	64.22	58.80	70.29	46.39	47.63	58.71	58.12	47.11	55.30
PRS[106]	61.56	47.81	33.09	66.61	60.13	70.89	47.45	44.71	60.55	58.15	54.39	68.92
AGCN	62.60	57.85	59.29	70.24	63.12	74.21	54.11	53.71	67.04	58.98	55.32	69.28
AGCN++	65.92	60.02	68.82	73.41	68.55	74.54	56.63	57.51	73.58	61.04	60.22	72.40

Note: We report three more important metrics (%) for multi-label classification after the whole data stream is seen once on the 8-way Split-COCO and 7-way Split-WIDE in both IL-MLCL and CL-MLCL scenarios.

As shown in Fig 5.6, our method performs better than other comparison methods in every task. These experiments can prove the effectiveness of AGCN++ from more angles.

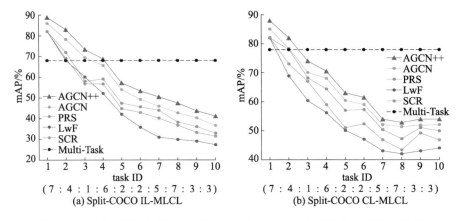

Fig. 5.6 mAP (%) of different classes setting in both IL-MLCL and CL-MLCL

5.3.7 mAP Curves

Similar to Bang, Kim, Yoo, et al.[17], Zhou, Ye, and Zhan[97], Mai, Li, Kim, et al.[109], we show the mAP trends of different methods in Fig. 5.7 for sequential learning. These curves indicate the performance along the MLCL progress. In two MLCL scenarios, Fig. 5.7 illustrates the mAP changes as tasks are being learned on two benchmarks. The mAP curves show that AGCN and AGCN++ can perform better through the MLCL process. In addition, their algorithm is applied after the first task for most continual learning methods. Our AGCN and AGCN++ has modelled the label dependencies from the first task. As distillation loss and relationship-preserving loss are applied to subsequent tasks, the algorithm's performance exceeds other methods in each task.

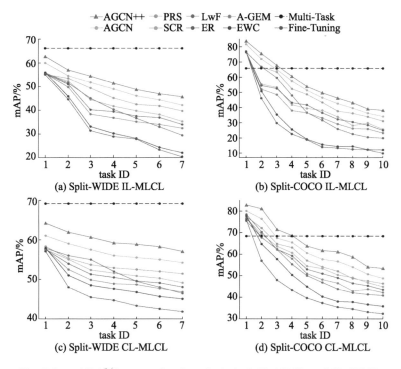

Fig. 5.7 mAP (%) on two benchmarks in both IL-MLCL and CL-MLCL

5.3.8 Ablation Studies

5.3.8.1 ACM and PLE effectiveness

We perform ablation experiments on ACM to test the effectiveness of the intra- and inter-task relationships for both AGCN and AGCN++. As shown in Sec. 5.2.5, R^t (Old-New) and Q^t (New-Old) are used to model inter-task label dependencies cross old and new tasks, B^t (New-New) is used to model intra-task label dependencies, and while R^t and Q^t are unavailable, neither are R^{t-1} and Q^{t-1} in block A^{t-1}, the A^{t-1} that inherit from the old task only build intra-task relationships. As shown in Table

5.5, if we do not build the label relationships across old and new tasks (w/o R^t & Q^t), the performance of AGCN is already better than most non-AGCN methods. For example, the comparison between AGCN (w/o R^t & Q^t) and LwF (w/o A^{t-1} & B^t and R^t & Q^t) on mAP is: 38.05% vs. 29.46% (Split-WIDE, IL-MLCL), 49.47% vs. 46.44% (Split-WIDE, CL-MLCL), 31.52% vs. 19.95% (Split-COCO, IL-MLCL) and 44.53% vs. 40.87% (Split-COCO, CL-MLCL), as shown in Table 5.5, Table 5.2 and Table 5.3. AGCN is also better than most non-AGCN methods on CF1 and OF1. This means only intra-task label relationships are effective for MLCL image recognition. When the inter-task block matrices $\mathcal{R}^t$ and $\mathcal{Q}^t$ are available, AGCN with both intra- and inter-task relationships (Line 2 and 4) can perform even better in all three metrics. For example, mAP comparisons of AGCN (w/A^{t-1} & B^t, R^t & Q^t) and PRS on two datasets in two scenarios: 42.15% vs. 39.70% (Split-WIDE, IL-MLCL), 54.20% vs. 51.42% (Split-WIDE, CL-MLCL), 34.11% vs. 31.08% (Split-COCO, IL-MLCL) and 48.82% vs. 46.39% (Split-WIDE, CL-MLCL), as shown in Table 5.5, Table 5.2 and Table 5.3, which means the inter-task relationships can enhance the multi-label recognition.

Table 5.5 Ablation studies (%) for ACM A^t and PLE on Split-WIDE and Split-COCO

			Split-WIDE						Split-COCO					
	A^{t-1} & B^t	R^t & Q^t	AGCN++ (w/PLE)			AGCN (w/o PLE)			AGCN++ (w/PLE)			AGCN (w/o PLE)		
			mAP↑	CF1↑	OF1↑	mAP↑	CF1↑	OF1↑	mAP↑	CF1↑	OF1↑	mAP↑	CF1↑	OF1↑
IL-MLCL	✓	✗	42.25	40.47	42.98	38.05	34.03	42.71	35.19	39.98	37.62	31.52	30.37	34.87
	✓	✓	45.73	43.04	45.26	42.15	37.99	43.70	38.23	41.38	45.26	34.11	35.49	42.37
CL-MLCL	✓	✗	54.72	50.41	48.84	49.47	44.73	52.13	51.51	47.13	56.97	44.53	35.55	53.57
	✓	✓	57.07	54.66	59.29	54.20	46.13	55.35	53.49	49.55	59.32	48.82	39.18	56.76

And AGCN++ (w/ PLE) outperforms AGCN (w/o PLE) either (w/o R^t & Q^t) or (w/A^{t-1} & B^t, R^t & Q^t) in all three metrics, which can prove the effectiveness of PLE. When w/o R^t & Q^t, for mAP, 42.25% vs. 38.05% (Split-WIDE, IL-MLCL), 54.72% vs. 49.47% (Split-WIDE, CL-MLCL), 35.19% vs. 31.52% (Split-COCO, IL-MLCL), 51.51% vs. 44.53%(Split-COCO, CL-MLCL). When w/ A^{t-1} & B^t, R^t & Q^t, for mAP, 45.73% vs. 42.15% (Split-WIDE, IL-MLCL), 57.07% vs. 54.20% (Split-WIDE, CL-MLCL), 38.23% vs. 34.11% (Split-COCO, IL-MLCL), 53.49% vs. 48.82% (Split-COCO, CL-MLCL).

5.3.8.2 Hyperparameter selection

Then, we analyze the influences of loss weights and relationship-preserving loss on two benchmarks, as shown in Table 5.6. When the relationship-preserving loss is unavailable, loss weight λ_3 is set to 0. The loss weights of others: $\lambda_1=0.10$, $\lambda_2=0.90$ for Split-WIDE in IL-MLCL, $\lambda_1=0.70$, $\lambda_2=0.30$ for Split-WIDE in CL-MLCL, $\lambda_1=0.15$, $\lambda_2=0.85$ for Split-COCO in ILMLCL and $\lambda_1=0.40$, $\lambda_2=0.60$ for Split-COCO in CL-MLCL. By adding the relationship-preserving loss l_{gph}, the performance gets more gains, and the values of forgetting are also lower, which means the mitigation of relationship-level catastrophic forgetting is quite essential for MLCL image recognition, and the relationship-preserving loss is effective. We select the best λ_3 as the hyper-parameters, i.e. $\lambda_3=10^4$ for Split-WIDE in IL-MLCL, $\lambda_3=10^3$ for Split-WIDE in CL-MLCL, $\lambda_3=10^4$ for Split-COCO in IL-MLCL and $\lambda_3=10^4$ for Split-COCO in CL-MLCL.

Table 5.6 Ablation studies (%) for loss weights and relationship-preserving loss on Split-WIDE and Split-COCO for IL-MLCL and CL-MLCL

	Split-WIDE						Split-COCO					
	λ_1	λ_2	λ_3	mAP↑	CF1↑	OF1↑	λ_1	λ_2	λ_3	mAP↑	CF1↑	OF1↑
IL-MLCL	0.10	0.90	0	42.04	38.76	42.12	0.15	0.85	0	36.77	39.95	39.10
	Forgetting↓			10.48	5.24	6.42	Forgetting↓			22.34	12.73	13.32
	0.10	0.90	10^4	45.73	43.04	45.26	0.15	0.85	10^4	38.23	41.38	45.26
	Forgetting↓			8.32	2.13	3.98	Forgetting↓			20.12	11.34	6.78
CL-MLCL	0.70	0.30	0	55.68	51.58	49.24	0.40	0.60	0	50.98	46.82	54.70
	Forgetting↓			5.04	12.56	10.24	Forgetting↓			12.87	22.80	6.62
	0.70	0.30	10^3	57.07	54.66	59.29	0.40	0.60	10^4	53.49	49.55	59.32
	Forgetting↓			4.45	10.64	1.02	Forgetting↓			7.24	14.52	1.24

5.3.9 Visualization of ACM

As shown in Fig. 5.8, to verify the effectiveness of the constructed

ACM, we offer the ACM visualizations on Split-WIDE and Split-COCO for IL-MLCL and CL-MLCL. We introduce the oracle augmented correlation matrix (oracle ACM) as the upper bound, which is constructed offline using hard label statistics of all tasks from corresponding datasets. d represents the Euclidean distance between the matrix and Oracle ACM. A smaller value of d means that the matrix is closer to Oracle ACM, which proves that this ACM is better constructed. As shown in Fig. 5.8, the proposed ACM in both scenarios is close to the oracle ACM. This indicates constructing ACM with soft or hard label statistics is effective. Note that in CL-MLCL, the ACM is constructed using only hard labels from the dataset, the ACMs of AGCN++ and AGCN are the same in the CL-MLCL scenario under the same dataset. And in IL-MLCL, the ACM is constructed with the soft labels produced by the model, so the ACM of AGCN++ is better constructed. In IL-MLCL, we can also observe that the ACM built in AGCN++ is closer to the Oracle ACM than in AGCN, 4.01% vs. 4.18% (Split-COCO) and 4.28% vs. 4.54% (Split-WIDE), which can prove that the PLE has reduced the accumulation of errors in the construction of label relationships.

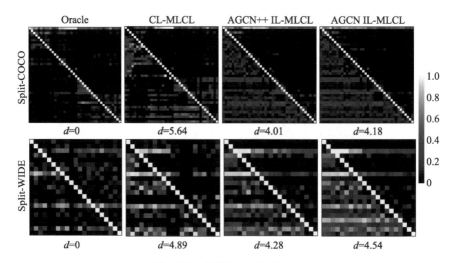

Fig. 5.8 ACM visualization

5.4 Chapter Conclusion

Multi-Label Continual Learning (MLCL) focuses on solving multi-label classification in continual learning. It is challenging to construct convincing label relationships and reduce forgetting in MLCL because of the partial label problem. This chapter proposed a novel AGCN++ based on an auto-updated expert mechanism to solve the problem of MLCL. The key of our AGCN++ is to construct label relationships in a partial label data stream and reduce catastrophic forgetting to improve overall performance. We studied MLCL in both IL-MLCL and CL-MLCL scenarios. In relationship construction, we showed the effectiveness of leveraging soft or hard label statistics to update the correlation matrix, even in the partial label data stream. We also showed the effectiveness of PLE in reducing the accumulation of errors in the construction of label relationships and suppressing forgetting. In terms of forgetting, we proposed an effective distillation loss and a novel relationship-preserving loss to mitigate class- and relationship level forgetting. Extensive experiments demonstrate that the proposed AGCN++ can capture well the label dependencies, thus achieving better MLCL performance in the IL-MLCL and CL-MLCL. Future work will study how to improve the construction of the old-old block using the correlation of only soft labels instead of inheriting the previously constructed ACM. It is believed that the performance will be further enhanced.

Chapter 6

Towards Long-Term Remembering for Federated Continual Learning

6.1 Introduction

In the era of big data, due to the popularity of edge devices such as smartphones and wearables, the privacy of individuals' data is important. In the case of private data distribution among a large number of edge devices[117], Federated Learning (FL)[118] does not require the transmission of raw data. Instead, the model in FL moves from the clients to the server in communication rounds[119-121].

However, classical FL cannot deal with incremental tasks and suffers from catastrophic forgetting. Continual Learning (CL)[27,28,40,105,122-127] continuously trains the model in a series of tasks of novel classes against forgetting[127] and is combined with Federated Learning, namely Federated Continual Learning (FCL)[128].

In FCL, clients learn from a sequence of tasks that can only be accessed privately, and the goal is to guarantee the server effectively learns novel knowledge while keeping performance on old tasks, i.e. reducing catastrophic forgetting. The existing solutions to FCL are divided into two categories. First, the distillation-based method[18,19] takes the server as the teacher model and uses public datasets to suppress forgetting. Second, the regulation-based method[128] constructs regularization on the server with server-only datasets.

However, the previous FCL methods can hardly retain earlier knowledge, that is, long-term memory cannot be kept well. Since the clients learn from different private datasets and cannot judge the co-important parameters in models, the server may indistinguishably mix up the uploaded parameters. The parameters regarding remembering are more likely to be overwritten, leading to catastrophic forgetting.

Thus, a key question of FCL is to answer how to obtain the co-important neurons of new tasks from different uploaded clients' models and avoid the overlap of the important neurons of new tasks and old tasks. Motivated by this, we propose Fisher INformation Accumulated Learning (FINAL) to retain long-term memories in Federated Continual Learning.

First, we evaluate the client-specific parameter importance via building the Fisher matrix following EWC[129] for each client, which is uploaded to the server independently. Then, we propose to identify the task-specific co-important neurons among clients through a Multi-node Collaborative Integration approach on the server to assemble a federated Fisher, which enables the integration of client-specific Fisher information on the current task. After clients train each task, the global Fisher absorbs the federated Fisher, capturing the important neurons for each task to retain long-term remembering.

After that, to avoid the overlap of the important neurons of new tasks and old tasks, we propose to distribute different Fisher combinations to clients and judge the optimal factors, namely Fisher balancing. By FINAL, we can maintain a global Fisher matrix to suppress long-term forgetting of FCL in an accumulated process. As shown in Fig. 6.1, FedAvg[121] suffers catastrophic forgetting and FedCL[127] cannot retain long-term knowledge on RainbowMNIST Tasks 1 to 4, especially on earlier tasks, while our FINAL can remember previous knowledge. We evaluate the proposed FINAL on four FCL datasets, and the results show that FINAL can reduce long-term forgetting and outperforms previous SOTA methods.

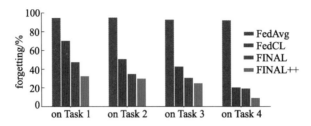

Fig. 6.1 The forgetting of old tasks of the final model after training 5 tasks on RainbowMNIST

6.1.1 Federated Learning

FedAvg[121], the most classic method, introduces the FL setting to machine learning. Based on FedAvg, the state-of-the-art methods for federated learning can be categorized into three main types. First, some methods[19,130] limit the updating direction of the local model and coordinate the optimization goal of the local model and the global model. FedProx[19] incorporated a proximal term in local model training. MOON[130] utilized the similarity between model representations to correct the local training of individual models. Second, some works are based on knowledge distillation via pseudo data to aggregate knowledge from clients to the server. For example, FedDF[131] proposed ensemble distillation for model fusion, training the central classifier through unlabeled data on the outputs of the models from the clients. FedFTG[132] fine-tuned the global model using pseudo data through knowledge distillation. Third, personalized federated learning has received wide attention in recent years. FedAMP[133] provided a way for each federated learning participant to learn personalized models. However, the FL algorithms are not suitable in the case of class-incremental scenarios. The server model will suffer from catastrophic forgetting.

6.1.2 Federated Continual Learning

In recent years, researchers have found that traditional federated learning cannot tackle changing data domains, i.e. new tasks or new classes, and they turn to study FCL, which is a brand new research direction. Federated Reconnaissance[124] introduces continual learning settings into the federal framework, making the server recognize the classes seen by clients. FedCL[127] evaluated importance weights on a small proxy dataset on the server to constrain local training. FLwF-2T[125] used the server as a second teacher to send its knowledge to a student client model. However, to protect privacy, FCL always saves no previous data from old tasks. The previous methods cannot guarantee good remembering of long-term tasks. Instead, in this chapter, we propose to accumulate clients' Fisher information to balance the old task remembering and the new task learning.

6.2 Methodology

6.2.1 Problem Definition

Given K clients in total, each client learns task by task in the same order from 1 to T. For the client k, we have a sequence of tasks and the corresponding training private datasets $\{\mathcal{D}_1^k, \cdots, \mathcal{D}_T^k\}$. For the Task t, after the clients complete their training, they upload their trained model $\{\theta_t^1, \cdots, \theta_t^k\}$ to ensemble a server model θ_t. The goal of FCL is to train a robust server model ensembling knowledge of all clients and all seen tasks without forgetting. That is, the final global server model is expected to predict all seen classes with the framework of FINAL, as shown in Fig. 6.2.

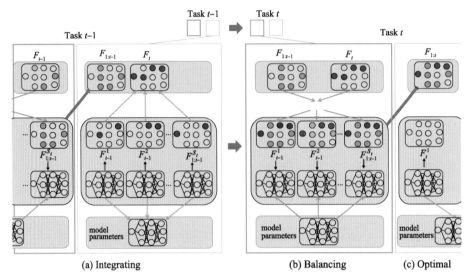

We use two colors to identify the upload and download processes.

Fig. 6.2 The framework of FINAL

6.2.2 Multi-Node Collaborative Integration for Parameter Co-Importance

Following previous works[126,128], in each round of communication, we randomly select S_t clients ($S_t < K$) for only one new task and transform FCL into a probability problem inspired by Santoro, Bartunov, Botvinick, et al.[134] Specifically, let $\mathcal{D}_t = \bigcup_{k<S_t} \mathcal{D}_t^k$ be the data gather of all clients. Suppose the model is now trained from Task $t-1$ to Task t, the server model can be obtained using Bayes' theorem:

$$p(\theta \mid \mathcal{D}_{1:t}) \propto p(\theta \mid \mathcal{D}_{1:t-1}) p(\mathcal{D}_t \mid \theta). \tag{6.1}$$

The posterior probability $p(\theta \mid \mathcal{D}_{1:t})$ contains the importance of the parameter learned from Task 1 to Task t. The true posterior of the server, however, is difficult to obtain for two reasons. (1) The private datasets of clients are unavailable for the server. (2) The past datasets of old tasks are unavailable for the server. The previous methods[125-127] introduce a public server-only dataset to reevaluate the posterior rather than the data from which the model was trained, which may contain large biases.

In this chapter, we propose to accumulate the Fisher from clients and previous tasks. To solve the first problem, we propose to approximate the posterior of each client via Laplace approximation, which fits the posterior by a Gaussian distribution of a task. For client k, we use Taylor Expansion

$$\log p(\theta | \mathcal{D}_t^k) \approx f(\theta_t^{k,*}) + \frac{1}{2}(\theta - \theta_t^{k,*})^2 f''(\theta_t^{k,*}), \tag{6.2}$$

where $f''(\theta_t^{k,*})$ is the Hessian of optimal solution $\theta_t^{k,*}$. The Gaussian is approximated with $\mathcal{N}\left(\theta_{1;t}^*, -\frac{1}{f''(\theta_{1;t}^*)}\right)$.

Because the Hessian is difficult to compute and store on edge devices, referring to EWC[134], we take the diagonal of Fisher information matrix $\boldsymbol{F}$ as an approximation of Hessian. A diagonal item F_{ii} is computed by

$$\boldsymbol{F}_{t,ii}^k \approx -\boldsymbol{E}_{(x,y) \sim \mathcal{D}_t^k} \nabla_{\theta_i}^2 \log p(y|x)|_{\theta = \theta_t^{k,*}}. \tag{6.3}$$

Fisher information evaluated in the client model retains the particularity of each model, that is, taking into account how much attention different models pay to the data. After training a task, all client-specific Fishers will be sent to the server without data leakage.

As shown in Fig. 6.2, each neuron in the Fisher has a different value, where a larger value indicates an important parameter. To merge the Fishers from multiple clients into a general task-specific Fisher, we propose a strategy called Multi-node Collaborative Integration. This strategy involves the collaborative integration of the uploaded Fishers as well as the parameters from different clients:

$$F_t = \sum_{k=1}^{S_t} I(\mathcal{D}_t^k) \cdot F_t^k, \theta_t = \sum_{k=1}^{S_t} I(\mathcal{D}_t^k) \cdot \theta_t^k. \tag{6.4}$$

Note we take the same integration strategy for Fishers and parameters to guarantee the consistency between models and important measures. It is open to select an integration strategy $I(\mathcal{D})$. We take a simple strategy that merges the Fisher value on each neuron by the scale of the dataset of each client:

$$I(\mathcal{D}_t^k) = \frac{|\mathcal{D}_t^k|}{\sum_{k=1}^{S_t} |\mathcal{D}_t^k|}. \tag{6.5}$$

The integrating Fisher information can retain the diversity of the current knowledge of each client on the task, focusing on more of the co-important parameters. To address this problem, we propose a method to accumulate Fisher information from F_1 to F_t and achieve $F_{1:t}$, enabling long-term remembering.

In other words, instead of storing and transmitting individual integrated Fishers, we introduce a mechanism to gradually accumulate and update the Fisher information over time. This accumulation process ensures that important information from earlier tasks is retained while incorporating new knowledge from subsequent tasks. By maintaining an accumulative Fisher matrix $F_{1:t}$, we achieve long-term memory retention without excessive communication overhead.

6.2.3 Fisher Accumulating and Balancing for Reducing Forgetting

In our proposed method, we update the client's optimal parameters for Task $t+1$ using the following objective function:

$$\theta_{t+1}^{k,*} = \arg\min_{\theta} L_{t+1}(\theta) + \frac{\lambda}{2} \sum_i F_{1:t,ii}^k (\theta_i - \theta_{t,i})^2. \tag{6.6}$$

where λ denotes the importance of the previous task following EWC. When learning a new task, the important parameters of the old task are constrained, and the optimal solution of the new task is restricted near the optimal solution of the old task. By accumulating the Fisher information matrix $F_{1:t}^k$, which accumulates information from all previous tasks, we can capture the long-term dependencies and retain important parameters for earlier tasks. This allows for better preservation of knowledge and reduces forgetting.

The accumulating Fisher information matrix $F_{1:t}^k$ continuously expands the evaluation of Fisher matrices for each task, which provides a comprehensive and integrated understanding of the importance of parameters across all tasks. Moreover, instead of rigidly assigning fixed weights or factors, our method dynamically incorporates the accumulated Fisher information from earlier tasks, effectively balancing the impact of new learning on existing knowledge. Every time the collaborative integrating Fisher information is calculated on a task, the global one will be updated by a weighted combination of the two Fisher information:

$$\begin{aligned}\boldsymbol{F}_{1:t} &= \eta_t \boldsymbol{F}_{1:t-1} \oplus \mu_t \boldsymbol{F}_t \\ &= \eta_t(\eta_{t-1}\boldsymbol{F}_{1:t-2} \oplus \mu_{t-1}\boldsymbol{F}_{t-1}) \oplus \mu_t \boldsymbol{F}_t \\ &= \cdots \\ &= \eta_t(\eta_{t-1}(\cdots(\eta_3 \boldsymbol{F}_{1:2} \oplus \mu_3 \boldsymbol{F}_3) \oplus \mu_t \boldsymbol{F}_t\end{aligned} \quad (6.7)$$

where $\eta_k + \mu_k = 1$ and η_k is the important factor of earlier tasks and $\oplus$ is the element-wise summation. Larger η means the global Fisher pays more attention to earlier knowledge.

Algorithm 5: Algorithm of FINAL++

ServerExecute:
1. for rounds(task) t in $\{1,2,\cdots,T\}$ do
2.   Randomly select S_t clients and sample $\{\eta^1,\cdots,\eta^{S_t}\}$;
3.   for client k in S_t do
4. $\boldsymbol{F}^k_{1:t-1} \leftarrow \eta^k \boldsymbol{F}_{1:t-2} + \mu^k \boldsymbol{F}_{t-1}$;
5. $\boldsymbol{F}^k_t, \theta^k_t \leftarrow$ Client Update($\boldsymbol{F}^k_{1:t-1}, \theta_{t-1}$) :: $\boldsymbol{F}^k_{1:t-1}, \theta_{t-1}$
6.   ;
7.   end
8.   Compute $\boldsymbol{F}_t, \theta_t$ and update θ_t by Eq.(6.4);
9. $\eta^* = \min(L_1, L_2, \cdots, L_k)$;
10. $\boldsymbol{F}^k_{1:t} \leftarrow \eta^* \boldsymbol{F}_{1:t-1} + \mu \boldsymbol{F}_t$
11. end

ClientUpdate($\boldsymbol{F}^k_{1:t-1}, \theta_{t-1}$) ::
12. Train local model θ^k_t using Eq.(6.6);
13. Estimate $\boldsymbol{F}^k_t$ by Eq.(6.3);
14. Compute loss by validation data: $L_k \leftarrow f(\theta^k_t, V_d)$;
15. return $\boldsymbol{F}^k_t, \theta^k_t$;

To address the conflict between old and new tasks when merging Fisher matrices, we adopt two communication-efficient strategies in our method. The first strategy is to set the η as a hyper-parameter to be searched, which balances the merging process. We name our method with this strategy as FINAL. The second strategy is to uniformly sample η from a discrete uniform distribution between 0 and 1 for each client, denoted as η_k. Each client k trains validation dataset[126-128] to evaluate and determine which client will be used. We introduce a public server-only using the respective Fisher matrices $F_{1:t-1}^k$ as a better combination. We name it with this strategy as FINAL++.

In comparison with FINAL, FINAL++ combines Fisher information matrices with different important factors for different clients to evaluate the conflict between $F_{1:t-1}$ and F_t and find the optimal balance. Through accumulating Fisher, we keep the important parameters for the early tasks to achieve long-term knowledge preservation. We elaborate on our proposed approach in Algorithm 5.

6.3 Experiments

6.3.1 Experiment Details

6.3.1.1 Dataset

We use four datasets, including RainbowMNIST, CIFAR-10, CIFAR-100, and TinyImageNet200 in our experiments as shown in Table 6.1. We randomly select 1% of the training data from each task as the task-incremental server-only validation data, and the rest 99% are used for all compared methods' training.

Table 6.1 The details of four datasets

Dataset	# Class	# Train	# Ser-Val
RainbowMNIST	2 classes × 5 tasks	59,400	600
CIFAR-10	2 classes × 5 tasks	49,500	500
CIFAR-100	10 classes × 10 tasks	49,500	500
TinyImageNet200	20 classes × 10 tasks	99,000	1,000

6.3.1.2 Implementation details

We set $K=10$ clients in total for the four datasets. In each communication round at Task t, we randomly select $S_t=5$ clients from the K clients following the FL setting. We use resnet-18 as the training model. The last classification layer of the model is incremented by the number of training classes. We use Adam optimizer, with a learning rate of 0.0001 on

RainbowMNIST and TinyImageNet200 and 0.001 on CIFAR-10 and CIFAR-100. The size of the mini-batch is 32 for all datasets. In our experiments, we set up 10 random fixed seeds, and the final result is the average of the 10 seeds.

6.3.1.3 Evaluation metric

We evaluate the performance via two common metrics. (1) Accuracy: the fraction of test samples predicted by the server model. (2) Forgetting: the subtraction of accuracy of the current training task model and the final model.

6.3.2 Results

6.3.2.1 Major comparisons

The major results are shown in Table 6.2, which shows the accuracy and forgetting of all compared methods averaged under 10 different random seeds. We evaluate a situation in which we adopt a fixed factor by our experience in Fisher balancing, shown as FINAL. FINAL and FINAL++ show the best performance on 4 datasets than others. FedAvg[121], FairFed[105] and MOON[130] almost forget all the old knowledge. And other compared FCL methods[40,124-127] show certain advantages to FedAvg. The forgetting in FedAvg is large, while FINAL and FINAL++ can inhibit forgetting to a large extent, especially long-term forgetting.

Table 6.2 The final model's accuracy and forgetting of methods on every task (%, avg±std)

Method	RainbowMNIST		CIFAR-10		CIFAR-100		TinyImageNet	
	accuracy ↑	forgetting ↓	accuracy ↑	forgetting ↓	accuracy ↑	forgetting ↓	accuracy ↑	forgetting ↓
FedAvg[121]	20.29±0.44	77.63±0.37	18.75±0.30	69.21±2.76	7.63±0.14	58.90±3.46	6.07±0.12	52.89±0.97
SI[57]	33.17±0.83	42.13±2.71	24.22±1.20	31.48±2.85	16.53±0.44	29.46±2.35	12.91±0.38	25.37±1.08
MAS[123]	34.66±0.36	34.19±1.02	25.09±0.57	29.42±2.31	17.37±0.63	27.91±1.78	13.60±0.42	24.81±1.54
FLwF-2T[125]	31.49±0.24	42.54±1.99	23.07±1.18	36.15±2.74	18.13±0.72	27.58±2.16	12.38±0.75	21.04±1.17
FedCL[18]	33.09±0.95	43.45±2.48	23.75±1.49	30.28±1.95	16.17±0.52	29.60±1.84	11.47±1.02	21.35±1.33
FCCL[135]	36.34±1.05	28.72±1.45	26.84±1.31	31.48±1.48	18.26±0.73	29.07±1.73	13.78±0.57	20.78±0.96

continued

Method	RainbowMNIST		CIFAR-10		CIFAR-100		TinyImageNet	
	accuracy ↑	forgetting ↓	accuracy ↑	forgetting ↓	accuracy ↑	forgetting ↓	accuracy ↑	forgetting ↓
MOON[130]	24.20± 0.46	72.52± 0.17	21.87± 0.53	65.23± 1.83	9.96± 0.26	55.93± 2.86	8.87± 0.22	50.55± 1.19
FairFed[136]	26.20± 0.33	66.79± 0.46	22.73± 0.61	57± 2.23	11.96± 0.18	50.93± 1.34	9.80± 0.20	43.55± 0.85
FINAL	39.45± 0.47	25.34± 1.58	31.81± 1.24	29.28± 2.33	20.13± 1.26	27.14± 1.84	15.95± 0.54	17.32± 0.94
FINAL++	51.61± 0.57	13.89± 2.46	39.70± 1.66	25.17± 1.90	24.69± 1.33	26.86± 1.45	17.89± 1.19	14.44± 1.01

6.3.2.2 Performance trend

As shown in Fig. 6.3(a), we show the performance trend of all compared methods on RainbowMNIST with the average performance after learning Tasks 1 to 5. Other methods show obvious forgetting on Task 1 while our FINAL has only a small range of forgetting. FINAL++ can maintain well on long-term memory and show the best performance because FINAL++ not only preserves the old knowledge but balances it with the new-trained task.

6.3.2.3 Evaluation on long-term remembering

Fig. 6.3(b) shows the final performance on all tasks after training with our FINAL++ and other state-of-the-art methods. We observe that FINAL can maintain long-term memory, while others cannot. In specific, the compared methods almost forget most of the knowledge of the earlier Tasks 1 to 3, resulting in poor performance.

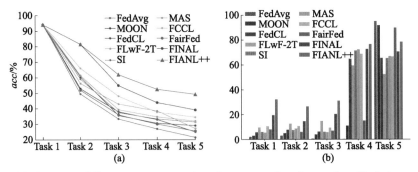

(a) Server model's accuracy on testing classes seen by clients after Task 1 to Task t. (b) Performance of long-term memory. The bar represents the final server model on every task class of different methods.

Fig. 6.3 Performance of the server model

6.3.3 Ablation Experiment

In this section, we study the importance of the major components of FINAL. In Table 6.3, we compare with the different uses of Fisher. First, we update without Fisher, i.e. "N/A". The results show much forgetting of old tasks. Then, we consider taking fixed η with Fisher integrating and Fisher balancing. Easy to see, that the higher the η, the higher the focus on the earlier tasks. It is worth mentioning that when $\eta = 0.0$, FINAL is only the integrating module without balancing. Another finding is that the fixed combination between Fishers is not practical in FCL, because the conflict between tasks is ignored. So the balance of the global Fisher and the new integrating Fisher is effective.

Table 6.3 The importance of Fisher balancing

Unit: %

η	N/A	0.0	0.1	0.2	0.3	0.4	0.5	FINAL++
Task 1	0.00	1.10	5.50	3.10	15.10	18.00	23.10	19.40
Task 2	0.40	1.60	9.90	10.29	18.30	18.60	24.40	17.20
Task 3	0.20	0.80	5.20	7.20	8.60	9.60	6.80	14.90
Task 4	0.30	1.50	12.50	14.50	19.69	17.20	18.60	21.60
Task 5	0.20	2.40	12.60	11.39	9.10	8.90	6.40	13.59
Task 6	0.00	7.86	19.20	17.10	16.60	13.70	17.50	17.10
Task 7	0.30	17.70	16.89	12.30	17.70	11.70	17.70	17.20
Task 8	0.00	17.90	24.20	16.29	19.80	15.90	14.90	14.29
Task 9	0.00	25.40	17.79	19.60	14.60	14.40	13.60	20.90
Task 10	58.80	28.10	21.70	21.20	7.20	13.50	7.20	22.60
Avg	6.02	10.44	14.55	13.30	14.67	14.15	15.02	17.89

Note: "N/A" represents no Fisher, and then η is set to from 0.0 to 0.9, meaning the fixed combination fraction of $\boldsymbol{F}_{1:t-1}$ and $\boldsymbol{F}_t$. The performance is evaluated on the final model in every task's classes.

We also study the validation data in Table 6.4 by using different datasets as validation. We can see that validation and training data show the best when they are in the same domain. Data of the same domain is optimal as validation. Although the effect of different domain datasets as validation decreases slightly, it still has clear improvement.

Table 6.4 The performance of different validation data

Train	RainbowMNIST	CIFAR-10	CIFAR-100	TinyImageNet
RainbowMNIST	51.61±0.57	49.21±0.51	49.52±0.43	49.05±0.72
CIFAR-10	37.96±1.59	39.70±1.66	38.32±1.48	38.73±1.25
CIFAR-100	23.69±1.11	24.03±1.51	24.69±1.33	13.72±1.27
TinyImageNet200	17.61±0.89	17.45±0.94	17.59±0.82	17.89±1.19

To explore the impact of different task sequences, we randomly scrambled 2 different task orders on RainbowMNIST to show the effects. Under the same setting, we get the results shown in Table 6.5. Through experiments, we find that our proposed method can still inhibit long-term forgetting on different task sequences in FCL and outperforms other methods. Besides, we set $\lambda = 1300$ in RainbowMNIST, and we give the ablation experiment of λ in Table 6.6.

Table 6.5 The accuracy and forgetting of methods on different task-sequence

Unit: %

Task-sequence	2,4,1,5,3		1,5,2,4,3		1,2,3,4,5 (in chapter)	
	accuracy ↑	forgetting ↓	accuracy ↑	forgetting ↓	accuracy ↑	forgetting ↓
FedAvg	21.65	76.31	20.93	67.57	20.29	77.63
SI	28.09	41.74	31.98	34.75	33.17	42.13
MAS	31.67	35.38	33.03	34.01	34.66	34.19
FLwF-2T	29.95	41.43	31.71	42.66	31.49	42.54
FedCL	32.13	44.4	33.07	46.38	33.09	43.45
FCCL	34.19	27.81	36.68	44.84	36.34	28.72
MOON	23.73	71.81	22.9	64.04	24.20	72.52
FairFed	25.27	66.72	25.59	59.89	27.14	66.94
FINAL	40.19	26.50	41.66	26.95	39.45	25.34
FINAL++	51.18	14.21	50.52	15.38	51.61	13.89

Table 6.6 Ablation experiment of λ

λ	900	1100	1300	1500	1700
accuracy	48.64	49.53	51.63	49.58	50.69

6.3.4 Fisher Visualization

Fig. 6.4 shows the difference of the Fisher on the last task with Fisher calculated at the beginning of the task. The higher value of the Fisher information means that the corresponding parameter is more important for the knowledge. By subtracting the new Fisher and the old one, we can see the forgetting (below 0) and learning (above 0) of methods. Without operation on reducing forgetting, FedAvg nearly loses all the important parameters. Other FCL methods also show significant forgetting of some parameters. In comparison, FINAL++ has less forgetting on neurons. In Fig. 6.4, we can confirm that our FINAL++ preserves and learns knowledge both effectively.

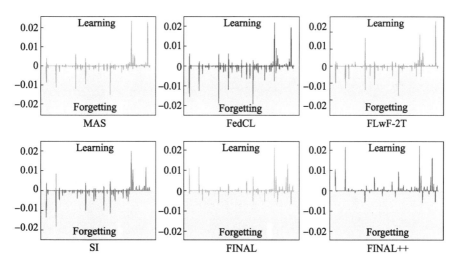

We compare the difference between the final Fisher information and the first Fisher information to show the preservation of knowledge. We use two colors to identify forgetting (below 0) and learning (above 0) separately.

Fig. 6.4 The difference between the first and last Fisher

6.4 Chapter Conclusion

FCL suffers from catastrophic forgetting, and the existing methods cannot maintain long-term remembering. In this chapter, we proposed a Fisher INformation Accumulation Learning (FINAL) method FCL. Our FINAL makes full use of the task-aware co-importance of each neuron in the network along the sequence of tasks from each client. For each client, we computed the Fishers and uploaded them to integrate for a task-specific Fisher, which denotes the parameter co-importance of this task. To balance the old and novel Fishers on the server, we leverage a Fisher balancing strategy, which finds the optimal important factor via distributing different Fisher combinations to Clients. The results show our FINAL can keep a long-term remembering.

Chapter 7

Centroid-based Rehearsal in Online Continual Learning

7.1 Introduction

Online Continual Learning (OCL) is used to enable a machine learning system to online learn from a sequence of tasks like humans[137], which has been applied to many applications, such as the recommender systems[7], medical research[138-140] and clinical management decisions[138]. However, OCL suffers from a well-known obstacle in neural networks called catastrophic forgetting[16,141,142], which is the inability to effectively retain old knowledge after learning a new task. The objective of OCL is to improve the adaptative ability to new knowledge over time without forgetting past knowledge.

To address the catastrophic forgetting, many methods have been proposed, such as regularization methods[143-145], Parameter-isolation methods[137,146,147] and rehearsal-based methods[35,43,44]. It is well-proven that the rehearsal-based method is simple and achieves impressive results in continual learning. However, as shown in Fig. 7.1(a), the traditional random sampling may select random samples with bias, and the following retraining on the biased data together with new tasks may lead to unpredictable shifts in feature space[145]. The phenomenon is named continual domain shifts[145], which makes forgetting even worse. In this chapter, we argue the cause of the continual domain shift is the stored data. Since the randomly-selected data points have a poor capacity to represent the original domain, retraining them will get over-fitting and indistinguishable. How to select data and make full use of the stored information become the key issues for rehearsal-based methods.

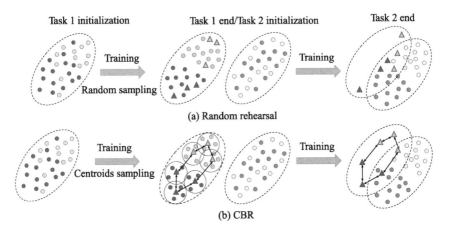

(a) Random sampling may select biased samples, which are prone to forget in the continual domain shift. (b) Our CBR selects diverse and representative data from centroids and suppresses continual domain shift through centroid distance distillation.

Fig. 7.1 Comparison between the random rehearsal and CBR in continual domain shift

Motivated by this, we propose a simple yet effective Centroid-Based Rehearsal (CBR) approach to mitigate the forgetting raised by continual domain shifting in OCL. A centroid is the online cluster center of the training domain[148-150]. Computing centroids for each task makes the model find the most representative location in the feature space, which contains the most comprehensive knowledge of tasks. However, implementing rehearsal needs to pick up the most representative samples among all training data, which is impractical in an online fashion[105,151,152]. To this end, we propose a centroid caching mechanism. The mechanism builds an auto-updated cache for each centroid for storing the most representative samples and the caches can also help the update of centroids. Moreover, we find that even the good centroids cannot preserve their knowledge well in

OCL because the relative relationship among centroids may shift together with the continual domain shift. To solve the problem, we propose to extra store the relative relationship, i.e. the pairwise distance among centroids. In the following training, we distill the centroid relationships on the memory buffer, which significantly retains the discriminative ability of the old task. As shown in Fig. 7.1(b), we build centroids for the old task, and the relative relationships among centroids are kept in the process of OCL. We demonstrate that our method can better alleviate the catastrophic forgetting raised by continual domain shift by the experiments on four popular OCL datasets. We summarize the contributions as follows:

(1) We propose a Centroid-Based Rehearsal (CBR) method to select samples by building and sampling from the auto-updated caches.

(2) We propose centroid distance distillation to constrain the centroid relationships.

(3) We evaluate CBR on four popular OCL datasets and obtain the new SOTA results.

7.2 Methodology

7.2.1 Continual Domain Shift in OCL

OCL can be formulated as learning from a sequence of datasets $\{\mathcal{D}_1, \cdots, \mathcal{D}_T\}$ in a data stream, where $\mathcal{D}_t = \{(x_i, y_i)\}_{i=1}^{N_t}$ is the data stream for the t-th task. We denote the model parameters as two main parts, the shared feature extraction layers $h: x \to f$ and the task-specific linear classification layer $g_t: f \to p$, where f and p are the deep feature and the predicted probability of the input image respectively.

In OCL, the model suffers from catastrophic forgetting due to the unavailability of data from past tasks. Rehearsal[153], as an effective method to reduce forgetting, stores a small number of samples, and the stored data will be retrained together with the current training. In specific, the memory is sampled online to fit $\mathcal{M} \sim \mathcal{D}^t$. During the new task learning, the episodic memory $\mathcal{M}$ is trained along with the new task data stream to achieve the effect of maintaining the past task knowledge. Although the continual learning method based on rehearsal help the ability of the model to remember past knowledge, the memory size is far small compared to the original data set $\mathcal{D}$, and the continual domain shift phenomenon in continual learning will inevitably occur.

Unfortunately, the existing sampling methods randomly fill the memory buffer, which may contain data bias such as outliers. Continual training on the biased data worse the continual domain shift. In this chapter, we propose a Centroid-Based Rehearsal (CBR) to preserve and make full use of as much comprehensive knowledge as possible, which can be seen in Fig. 7.2.

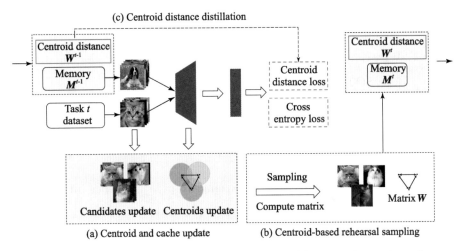

(a) The model computes a set of centroids and candidate samples at the cache layer. (b) Memory is selected from the cache by centroid-based rehearsal sampling. (c) We compute the centroid relationship of memory data and distill the stored relationship to reduce continual domain shift.

Fig. 7.2 The framework of Centroid-Based Rehearsal

7.2.2 Centroid-based Rehearsal

7.2.2.1 Centroid building

Offline continual learning methods have the whole dataset when learning new tasks, and it is easy to obtain the cluster using any clustering methods such as K-Means[87]. However, in OCL, we need to leverage the online clustering methods on data stream, i.e. centroid. Following Agg-Var[150], we update centroids with a given threshold ε. Given a new data point x, if the closest centroid c has a distance less than m, the centroid is updated via

$$c = \frac{n_c \times c + h(x)}{n_c + 1} \tag{7.1}$$

where n_c is the number of the data point (images) already represented by the centroid. Elsewise, a new centroid will be constructed using $h(x)$.

In the online fashion, we cannot pick up and store the representative sample in a fixed-size memory buffer for two reasons: (1) the samples of past and future are unavailable; (2) the centroid also makes shift with the training of model. To pick up representative samples for memory, we propose a centroid caching mechanism in the following section.

7.2.2.2 Centroid caching mechanism

We build centroid-aware caches as the medium to store the closest sample candidates to each centroid, and the cache will be discarded after the task is finished. For a centroid c, its cache is represented as $\mathcal{A}_c = \{(x_j, y_j)\}_{j=1}^{K}$. The candidate data are the K nearest data in the data stream to each centroid. The centroid caching mechanism contains two main procedures.

Centroid update. After the training of the current task, we select the sample data from the cache into memory and delete the data in the cache. To prevent bias in the centroids and the updated model, based on Agg-var [Eq. (7.1)], we formulate the centroid update as

$$c = \frac{\sum_{x' \in \mathcal{A}_c} h(x') + h(x)}{|A_c| + 1} \tag{7.2}$$

where c is the average of the features of the candidate data corresponding to this centroid (closest and distance less than ε). In Eq. (7.2), important centroids are updated more frequently than some outliers. Additionally, to reduce the impact of continual domain shift, we update the centroids with not only the new data but the corresponding caches. The caches are sent to the current model for updated features, which helps centroids precisely update.

Cache update. The centroid with the closest distance to the feature vector $h(x)$ is updated using Eq. (7.2). If the number of candidates for this centroid is less than the pre-defined limit γ, (x,y) is added to the cache directly. If the candidate number is equal to γ, the updated centroid compares the distance with $\gamma+1$ candidates and removes the farthest one.

The centroid caching mechanism builds a bridge between centroids and the memory buffer. On one hand, the cache helps the centroid more representative. On the other hand, we can select from the cache for more representative samples as illustrated in the next section.

7.2.2.3 Centroid-based rehearsal sampling

At any time during the training phase of Task t, we build a centroid-aware cache for each centroid in the previous section. Each centroid is updated via data points. In previous works, the sampling in OCL is always random, which may raise large data bias and be prone to shift in the feature space. The proposed centroid caching mechanism obtains the cluster centers online, and the cache of each centroid is regarded as the most representative data. Thus, we can sample data points directly from all caches to make up the task-specific memory buffer $\mathcal{M}^t$. However, the data distribution may contain outliers, where some centroids may be updated by only a few examples.

To reduce the influence of outliers, we propose to sample from caches according to the number of centroid updating. First, we select the samples from the cache based on the confidence level m_i, which is the total number of centroid updating. In general, the larger the m_i, the more representative the centroid it corresponds to. If m_i is small, its corresponding centroid is likely to be an outlier data point. Different centroids in the same class have different values. We convert m_i corresponding to each centroid to its probability of being selected. The memory buffer can be obtained by

$$\mathcal{M} = \bigcup_i^{|\mathcal{M}|} \{(x_i, y_i) \sim P^t((x, y))\}, \quad (7.3)$$

where $P_i^t = P^t((x, y) \in \mathcal{A}_{c_i}) = \dfrac{m_i}{\sum_{k=1}^{n} m_k}$. We use $P_i^t \in [0, 1]$ to represent the sampling probability of the i-th centroid. The larger the confidence m_i, the higher the sample probability of the corresponding cache being selected.

By changing the sampling probability of each cache, our method reduces the influence of outliers. It is worth noting why we do not only sample from caches with larger confidence because we also need to keep the diversity to some extent. Otherwise, the stored samples will have larger biases than random selection. Our centroid-based rehearsal sampling not only improves the representativeness of memory selection but also keeps the diversity.

Algorithm 6: Update centroid and cache

Input: Centroids $\{c_1, \cdots, c_N\}$, Centroid update numbers $\{K_1, \cdots, K_N\}$, model h, Cache $\mathcal{A}$, Distance threshold ε, Cache size γ, data (x, y)

Output: Updated $\{c_1, \cdots, c_{N'}\}$ and $\mathcal{A}$

1   ▽ Find nearest centroid;
2   $i^* = \mathrm{armin}_{i \in [1,N]} \| h(x) - c_i \|$;
3   $d = \| h(x) - c_{i^*} \|$;
4   if $d > \varepsilon$ then
5   ▽ Create new centroid;
6     $N' = N + 1$
7     $c_{N'} = h(x)$;
8   end
9   else
10   ▽ Update centroid;
11    $N' = N$;
12    $c_{i^*} = \dfrac{\sum_{j=1}^{K} h(x_j \in \mathcal{A}_{c_{i^*}}) + h(x)}{K_{i^*} + 1}$;
13   end
14   ▽ Update cache;
15   if $K_{i^*} < \gamma$ then
16    $\mathcal{A}_{c_{i^*}} \leftarrow \mathcal{A}_{c_{i^*}} \cup (x, y)$;
17   end
18   else
19    $j = \arg\max_j \| \mathcal{A}_{c_{i^*}} - c_{i^*} \|$;
20   Remove $\mathcal{A}_{c,j}$ from $\mathcal{A}_c$;
21   end

7.2.3 Distillation on Centroid Distance

Only selecting samples with diversity and representativeness for replay based on the centroids is not enough to effectively solve the continual domain shift problem. Because in the process of continual learning, domain shift will blur the decision boundary of the old task and cause forgetting of the old knowledge. A naive way is to anchor the memory feature as mentioned in previous works[154,155]. However, constraining the memory feature move may undermine the new task learning and the stored features are a large storage burden.

We argue that we do not have to store the memory feature due to the aim of OCL is only to keep the discriminative ability. Instead, we propose to only store the relative relationships among centroids, i.e. the Cosine distances. We build the optimization algorithm of centroids distance distillation. In specific, we calculate the distance between all the centroids of the current task (intra- and inter-class). Note that we do not directly use the online centroid in Eq. (7.2), we recalculate the cache centroid by averaging the features of the cache samples from the same centroid. Thus, the pairwise distances can be represented by a matrix W^t, where

$$W_{ij}^t = \frac{c_i \cdot c_j}{\|c_i\| \|c_j\|}, \text{where } c = \frac{1}{|\mathcal{A}_c|} \sum_{x \in \mathcal{A}_c} h(x). \tag{7.4}$$

The storage cost of this matrix is very small compared to the samples and their features.

In the learning of new tasks, we recompute the centroid distance matrix $W^{t'}$ using the memory and distill it with the stored matrix W^t with a Centroid Distillation (CD) loss

$$\mathscr{L}_{CD} = \sum_{t=1}^{k-1} \|W^t - W^{t'}\|^2. \tag{7.5}$$

The centroid distance matrix is distilled in the following tasks to keep the decision boundary from blurring. The CD loss takes advantage of the information between the centroids to further suppress the continual domain shift, where the diversity and representativeness will be kept even the continual domain shift occurs. The relative relationships among centroids also have the classes of old tasks discriminative from each other in the current tasks.

7.2.4 The Overall Algorithm

The main goal of OCL is to minimize the following objectives at Task t:

$$\mathcal{L}(\mathcal{B}) = \mathbf{E}_{(x,y)\sim\mathcal{B}} l(p, y), \mathcal{L}_{\text{reh}}(\mathcal{B}_{\text{reh}}) = \mathbf{E}_{(x,y)\sim\mathcal{B}_{\text{reh}}} l(p, y), \quad (7.6)$$

where $\mathcal{L}(.,.)$ represents cross-entropy loss. $\mathcal{B}$ represents the online minibatch at the moment. $\mathcal{B}_{\text{reh}}$ represents the rehearsal mini-batch in memory. However, relying solely on the above two loss functions cannot reduce domain shift. Our proposed $\mathcal{L}_{\text{CD}}$ keeps the feature domain stable by distilling the centroid distance. To further mitigate continual domain shift, following Abràmoff, Lavin, Birch, et al.[155], we use feature distillation loss as an optional term

$$\mathcal{L}_{\text{FD}} = \| h_\theta(x_{\text{reh}}) - f_{\text{reh}} \|^2 \quad (7.7)$$

where f_{reh} is the corresponding latent representations when memory data x_{reh} is first trained.

The whole objective of the proposed method is

$$\mathcal{L}_{\text{CBR}} = \mathcal{L} + \mathcal{L}_{\text{reh}} + \mathcal{L}_{\text{CD}} + \mathcal{L}_{\text{FD}} \quad (7.8)$$

where $\mathcal{L}$ focuses on learning the current task. $\mathcal{L}_{\text{reh}}$ helps the model recall knowledge from past tasks. $\mathcal{L}_{\text{CD}}$ and $\mathcal{L}_{\text{FD}}$ serve to mitigate the effect of continual domain shift in OCL.

Fig. 7.2 is the framework of our method. For each image in the data stream, the feature extractor generates features for each image and computes a set of centroids and candidate samples at the cache layer. Algorithm 6 summarizes the centroid building process. When the current task training is finished, a set of data is selected and stored in the memory buffer by centroid-based sampling. At each stage, the memory data recalculates the centroid distance matrix and distills the stored matrix to reduce continual domain shift in continual learning.

7.3 Experiments

7.3.1 Dataset and Experimental Details

7.3.1.1 Dataset

Permuted MNIST[140] is a variant of MNIST[146] dataset where the input pixels for each task have different random permutation as different tasks. Split CIFAR is a split version of the original CIFAR-100 dataset[156], which splits 100 classes into 20 disjoint tasks, each containing 5 classes. Split CUB is a random splitting of the 200 classes of the CUB dataset[58] into 20 disjoint tasks, each containing 10 classes. Split AWA is an incremental version of the AWA dataset[157] that splits 50 animal categories into 20 tasks, each with 5 different categories, but the categories are repeatable across tasks. We follow previous works[39,77,157] to conduct experiments on the above four datasets.

7.3.1.2 Evaluation metric

Average Accuracy (A_T) is the average of the accuracy of all tasks after the model is trained on the last task. Forgetting Measure (F_T) represents the the accuracy drop of old tasks after the model is trained on all tasks. Long-Term Remembering (LTR) computes the accuracy drop of each task relative to when the task was first trained.

7.3.1.3 Implementation details

Following previous works[60,158], we implement our methods with different backbones. For Permuted MNIST, we use a fully connected network with two hidden layers of 256 RELU units. For Split CIFAR, we use a reduced resnet18[12] network. For Split CUB and Split AWA, we use a standard resnet18 network. All networks employ ReLU in the hidden layers and the model is optimized using stochastic gradient descent with a mini-batch size of 10. For Permuted MNIST, Split CIFAR, Split CUB, Split AWA, ϵ and γ are respectively set to (7,35), (8,20), (11,10), (7,35).

7.3.2 Experimental Results

We compare the proposed CBR with multiple baseline methods in the CL setup. Regularization-based methods: EWC[18], MAS[128] and PI[161]. Regularization-based methods: GEM[43], A-GEM[35], ER[87], MEGA[46], DER[69], MDMT-R[29], ASER[160].

7.3.2.1 Main comparisons

As shown in Table 7.1, we compare CBR with other SOTAs in three metrics. For (A_T), our method shows a clear improvement compared to other methods on all four datasets. This indicates that our method can better suppress forgetting and reduce domain shift. For F_T and LTR, our method also has an advantageous position, which indicates that our method can gain long-term memory by reducing the domain shift. CBR achieves the best results compared to SOTAs without FD loss. With FD loss, CBR can be further improved.

Table 7.1 Comparisons on four datasets

Methods	Permuted MNIST			Split CIFAR		
	$A_T/\%$	F_T	LCR	$A_T/\%$	F_T	LCR
EWC[18]	68.68±0.98	0.28±0.010	3.292±0.135	42.67±4.24	0.26±0.039	2.493±0.427
A-GEM[35]	89.32±0.46	0.07±0.004	0.367±0.013	61.28±1.88	0.09±0.018	0.643±0.124
ER[87]	90.47±0.14	0.03±0.001	0.367±0.013	63.97±1.30	0.06±0.006	0.451±0.333
MEGA[46]	54.28±4.84	0.05±0.040	0.070±0.114	66.12±1.94	0.06±0.015	0.356±0.114
DER[159]	92.03±0.19	0.04±0.001	0.402±0.012	68.49±1.45	0.06±0.009	0.371±0.087
ASER[160]	—	—	—	65.53±1.89	0.07±0.007	0.544±0.133
SCR[109]	91.74±0.63	0.05±0.004	0.492±0.041	67.99±1.89	0.05±0.004	0.258±0.024
MDMT-R[29]	91.97±0.23	0.05±0.002	0.521±0.022	66.38±1.63	0.05±0.006	0.377±0.076
MDMTR+FD[29]	93.97±0.15	0.03±0.002	0.283±0.019	69.20±1.60	0.04±0.010	0.283±0.099
CBR	92.22±0.22	0.04±0.002	0.484±0.019	69.65±1.55	0.04±0.013	0.192±0.094
CBR+FD	94.12±0.11	0.01±0.007	0.013±0.009	70.69±2.33	0.03±0.027	0.119±0.065

continued

Methods	Split CUB			Split AWA		
	$A_T/\%$	F_T	LTR	$A_T/\%$	F_T	LTR
EWC[18]	53.56±1.67	0.14±0.024	1.021±0.210	33.43±3.07	0.08±0.021	0.675±0.214
A-GEM[35]	61.82±3.72	0.08±0.021	0.456±0.174	44.95±2.97	0.05±0.014	0.178±0.082
ER[87]	73.63±0.52	0.01±0.005	0.001±0.001	53.27±4.05	0.02±0.030	0.014±0.015
MEGA[46]	80.58±1.94	0.01±0.017	0.002±0.002	54.28±4.84	0.05±0.040	0.070±0.114
DER[159]	76.56±2.48	0.01±0.015	0.025±0.018	50.70±4.91	0.04±0.040	0.063±0.094
ASER[160]	75.58±3.72	0.02±0.010	0.037±0.029	46.72±3.20	0.05±0.006	0.171±0.021
SCR[109]	81.43±1.97	0.01±0.007	0.007±0.009	54.35±2.68	0.02±0.012	0.022±0.010
MDMT-R[29]	83.06±4.39	0.20±0.028	0.015±0.023	58.20±2.51	0.02±0.011	0.035±0.025
MDMTR+FD[29]	83.98±2.35	0.01±0.015	0.021±0.018	61.26±3.36	0.02±0.027	0.002±0.002
CBR	84.85±2.46	0.01±0.012	0.008±0.010	59.26±4.72	0.03±0.037	0.034±0.065
CBR+FD	85.75±1.99	0.01±0.004	0.004±0.006	61.92±2.94	0.02±0.027	0.013±0.002

Note: The mean and Std are over 5 seeds.

7.3.2.2 Ablation study

We then explore the impact of each CBR configuration. We show the ablation experiments on the Split CIFAR dataset in Table 7.2. The first row of the table shows the performance of our experimental baseline, without adding any configuration. When we add CD loss or FD loss, the experimental results can be clearly improved. The experimental performance is further improved when CD loss and FD loss are used together. We also conducted an experimental study on the choice of the hyper-parameter centroid distance ε. The larger the centroid distance, the less the class trains out of the centroid, and the smaller the centroid distance, the more the centroid is obtained. Easy to observe, when a suitable centroid distance is chosen, the experimental effect can reach optimal performance.

Table 7.2 Ablation study on Split CIFAR

ε	CD	FD	$A_T/\%$	F_T	LTR
—	—	—	66.38±1.63	0.052±0.006	0.377±0.076
—	—	✓	68.97±2.21	0.040±0.009	0.254±0.031

continued

ε	CD	FD	$A_T/\%$	F_T	LTR
8	✓	—	69.65±1.55	0.035±0.013	0.192±0.094
7	✓	✓	70.16±1.96	0.029±0.012	0.148±0.072
7.5	✓	✓	70.57±2.45	0.029±0.009	0.121±0.055
8.5	✓	✓	70.30±2.61	0.032±0.011	0.154±0.052
9	✓	✓	69.81±1.76	0.030±0.008	0.143±0.048
8	✓	✓	70.69±2.33	0.027±0.012	0.119±0.065

7.3.2.3 Cache size analysis

In Fig. 7.3, we also set three memory sizes on the Split CIFAR dataset to explore the effect of changing the cache size γ on the experimental results. We can see that for different memory sizes, the optimal cache sizes are different. Moreover, a larger memory size always means better performance with an appropriate cache size. In our implementation, we set different γ to different datasets.

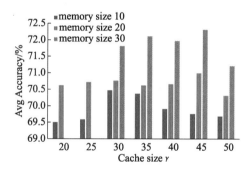

Fig. 7.3 The effect of the cache size γ with different memory size settings

7.3.2.4 Sampling strategy comparisons

In Fig. 7.4, we compare with other sampling methods. Ring buffer and reservoir sampling are the most classical online random sampling methods. Mean-of-Feature (MoF)[109] samples the data closest to the mean by calculating the mean of the features for all data in each class, which means larger storage is needed. In contrast, our centroid-based sampling achieves the best results with each metric without much storage.

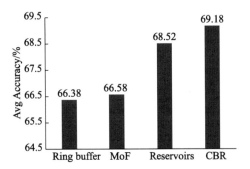

Fig. 7.4 Online sampling methods results on Split CIFAR

7.3.2.5 Continual domain shift observation

We explore the continual domain shift phenomenon in continual learning by the t-SNE visualization. In Fig. 7.5, we show the features distribution of Task 1 on Permuted MNIST at the end of Task 1 phase and Task 6 phase. The DER and SCR methods do not suppress the occurrence of continual domain shift and thus lead to the forgetting of past knowledge. CBR can effectively mitigate the continual domain shift by centroid-based sampling and centroid distance distillation, and the feature distribution remains relatively stable after the end of Task 6. When the CBR method is combined with FD loss, the continual domain shift phenomenon is further reduced.

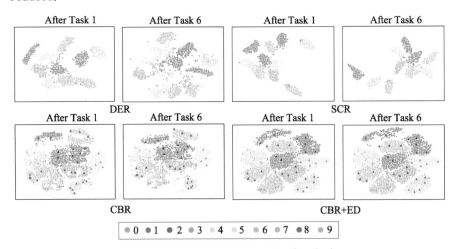

The darker points are the centroids of each class.

Fig. 7.5 Feature t-SNE of Task 1 on Permuted MNIST after the learning of Tasks 1 and 6

7.4 Chapter Conclusion

In this chapter, we address catastrophic forgetting, a major challenge of OCL study, by considering the continual domain drift of old tasks in the training sequence. In rehearsal based OCL, the continual domain drift is given rise from the imbalance of stored old tasks data and massive new tasks data. We propose an effective Drift-Reducing Rehearsal (DRR) method, which effectively anchors the domain of old tasks and makes all tasks perceive each other. First, we store samples in the memory buffer of rehearsal by measuring the distance from the sample to constructed centroids. Then, we propose a new cross-task contrastive margin loss to encourage intra-class and intra-task compactness, and inter-class and inter-task discrepancy. Finally, we present an optional centroid distillation loss to mitigate the continual domain drift. We evaluate the proposed DRR method on four OCL benchmark datasets. Extensive experiments show the superiority of our approach over SOTA methods. Although the method proposed in this chapter has achieved certain effectiveness, we do not have a way to constrain the exact direction of domain drift, making the direction of drift easily uncontrollable after training on a large number of tasks. In the future, we plan to research modeling the direction of domain drift and controlling the orthogonality of multi-task directions to ensure better OCL effectiveness.

Chapter 8

Dynamic V2X Perception from Road-to-Vehicle Vision

8.1 Introduction

Environmental perception[162] is important for autonomous vehicles[163], enabling them to express, identify, and interpret sensory inputs to understand complex environments. In recent years, there have been in-depth researches on single-vehicle perception[164], with researchers proposing various algorithms for handling different downstream tasks. Despite progress in single vehicle perception, individual perspectives often lead to a decline in perception capability in distant or obstructed areas. Vehicle-to-everything (V2X) technology[165-167] is an advanced communication system facilitating dialogue between vehicles and other entities[168,169], including vehicle-to-vehicle and vehicle-to-infrastructure communication, single-vehicle perception can be upgraded to cooperative perception. V2X communication introduces additional perspectives that assist autonomous vehicles in seeing farther, clearer, and even subtler obstacles, thereby enhancing perception capabilities.

While current V2X methods[170,171] showcase commendable perception capabilities in autonomous driving on straight roads, its performance falters in complex traffic intersections precisely the locations where traffic accidents are prone to occur. The challenging task in real-world environmental perception of traffic intersections comprises two key aspects: (1) Intra-scene variations: this involves the variations within a scene, containing factors like pedestrians in motion and vehicles executing turns. (2) Inter-scene variations: this pertains to substantial changes in architecture and lane configurations between different scenes. We argue that the reason for the failure of the current V2X methods in complex intersections is rooted in the over-reliance on the vehicle perception vision. Only aggregating vehicles in unpredictable locations may overlook some important scene elements around intersections, such as the pedestrians crossing streets. In general, these methods lack the effective use of static and stable visual information from the RSU. On top of this, the RSU[172] serves as an important support for the V2X system, because a stationary RSU positioned at a fixed location, provides a more stable and expansive sensory coverage while minimizing redundant communication. RSU facilitates the collection of comprehensive perception data over a wider range and ensures more consistent perception as compared to inherently mobile vehicles. Furthermore, RSU can exploit the increased storage and computing capabilities, to adapt to the wide range of inter-scene variations.

Nevertheless, training V2X using RSU to achieve effective perception in complex dynamic intersections is not a trivial issue. The primary challenge lies in how to adapt to the aforementioned two dynamic scenarios. On one hand, the locations of smart vehicles at each intersection are always uncertain, making it difficult to effectively integrate vehicle perception. The dynamicity of vehicles often complicates it even further to evaluate the performance of a V2X system[173]. As the positions and directions of vehicles constantly change, it is challenging to predict how they will affect traffic flow and safety. To achieve comprehensive perceptions of the dynamic traffic environment, it is necessary to collect data from vehicles at different angles and positions and analyze their data together to form a complete understanding of the entire intersection. On the other hand, existing have not fully considered the practical cross-scene learning process, resulting in perception models frequently experiencing abrupt errors in this process and lacking adaptability. In different traffic scenarios, such as busy city center intersections and suburban intersections with less traffic, perception models need to be able to adapt to these different environments and conditions. In this chapter, we focus on optimizing multi-perspective perception fusion within scenes and addressing cross-scene adaptability issues at intersections with the use of RSU, as existing technologies frequently struggle with transitioning between these scenarios, resulting in perception errors or failures.

Motivated by this, this chapter proposes an Adaptive Road-to-Vehicle Perception (AR2VP) approach, which constructs a road-to-vehicle cooperative perception model tailored for dynamic environments. As shown in Fig. 8.1, within dynamic traffic scenarios, RSU serves as the perception nodes situated at the heart of the traffic scene, equipped with additional storage space and computational capabilities. The geographical advantage of these units allows them to process and analyze traffic data more effectively. In each specific traffic scenario, RSU shows outstanding sensing capabilities, enabling them to capture and analyze the complexity of traffic flow from multiple perspectives. Our work re-explores the latent advantages of the RSU and leverages the road-to-vehicle vision paradigm to enhance V2X perception within dynamic scenes. First, to effectively handle intra-scene changes, we design DPR module and R2VPC module. These modules collaboratively construct an adaptable graph catering to intra-scene changes, thereby enhancing vehicles' overall adaptability within dynamic scenarios. Second, to effectively adapt to inter-scene changes, we present RSU-Prompt Replay (RPR), utilizing RSU limited storage capacity and typical sample replay techniques in continual learning[173,174], enabling vehicles to adapt to large-scale scene transitions beyond intra-scene changes. Our approach is validated on the tasks of scene segmentation and 3D object detection. The results on the V2XSim dataset[175] show good perception performance and the adaptability of AR2VP to dynamic scenes.

Chapter 8
Dynamic V2X Perception from Road-to-Vehicle Vision

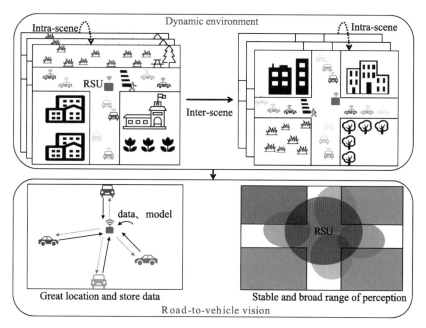

RSU demonstrates stability in perception and possesses geographical advantages in intra-scene changes, as opposed to the continual mobility of vehicles in inter-scene dynamics. With storage and communication capabilities, RSU stores old scenes data and model, enabling adaptation to inter-scene changes.

Fig. 8.1 Dynamic scene advantage of RSU

Our contributions are three-fold:

(1) We propose an AR2VP approach for dynamic V2X perception to address V2X perception challenges at complex traffic intersections in dynamic scenarios. AR2VP effectively coordinates perceptual information between vehicles and RSU at traffic intersections and applies to optimizing perception across different traffic intersections.

(2) We design DPR module and R2VPC module to deal with intra-scene changes. DPR leverages the geographical advantage and perceptual insights from roadside locations to establish an effective fusion of vehicle-road information. Meanwhile, R2VPC enhances the overall adaptability of vehicles in dynamic scenarios by improving the stability of roadside perception.

(3) We present RPR strategy to effectively handle inter-scene changes. Prompt-replay utilizes the perceptual completeness of updated RSU features to train learnable scene prompts, summarizing information about the entire intersection scene. This allows for the extraction of a small number of representative samples that characterize the intersection, enabling high model robustness with a limited RSU storage capacity.

V2X technology encompasses various forms of cooperative communication, including Vehicle-to-Vehicle (V2V)[166,167] and Vehicle-to-Infrastructure (V2I)[168]. All agents engage in mutual communication on the broadcast platform. The core step in V2X perception technology involves the adoption of a cooperative strategy. Current approaches mainly include raw measurement-based early collaboration, output-based late collaboration, and feature-based intermediate fusion.

Early fusion[171] integrates raw measurements from all agents, providing a comprehensive perspective to effectively address occlusion and long-range issues in single-agent perception. However, it requires substantial communication bandwidth. Late fusion combines each agent's perception outputs, emphasizing bandwidth efficiency. Yet, this approach may result in noisy and incomplete individual perception outputs, leading to suboptimal fusion results. To strike a balance between performance and bandwidth, intermediate Fusion[176,177] is introduced. This strategy aggregates intermediate features across agents[178], potentially achieving both communication bandwidth efficiency and enhanced perception ability. Nevertheless, insufficient collaboration strategy design may lead to information loss during feature abstraction and fusion, limiting improvements in perception ability.

For intermediate fusion technology, Who2com[170] utilizes a handshake communication mechanism to determine which two agents should be used for image segmentation, providing a simple yet effective communication approach. However, this method may be limited by considering communication only between two agents and may not be flexible enough in larger-scale scenarios. When2com[169] introduces an asymmetric attention mechanism to decide when to communicate and how to create communication groups for image segmentation. This method makes communication between agents more flexible, allowing for automatic adjustment of communication frequency as needed. However, due to the introduction of more parameters and complexity, it may require more computational resources and longer training times in practical applications. V2VNet[166] proposes multi-round message passing on a spatial-aware graph neural network for joint perception and prediction in autonomous driving. This method fully leverages the spatial relationships between vehicles and enables end-to-end learning in perception and prediction tasks, thereby improving overall performance. However, due to the need for larger computational overhead and more data to train complex neural network models, it may require more time and resources. DiscoNet[179] introduces a distilled collaboration graph with matrix-valued edge weights for adaptive perception, offering a superior performance-bandwidth trade-off. This method reduces communication overhead by using matrix-valued edge weights, thus reducing bandwidth requirements while maintaining high performance. However, the need for graph simplification and optimization may increase some computational complexity and training costs. V2I-CARLA[180] proposes a pose error regression module to learn to correct pose errors when the pose information from other vehicles is noisy. For V2I[180] technology, the collaboration is between infrastructure and vehicles, which expands the vehicle's perception field. Overall, these methods have their strengths and weaknesses. Who2com is straightforward but may be limited by small-scale communication. When2com algorithm offers greater flexibility but may require more computational resources. V2VNet maximizes spatial information but may require more data and time for training. DiscoNet provides a better balance between performance and bandwidth but may require more computational complexity. Compared to earlier research, these intermediate collaboration algorithms propose new methods for handling communication and cooperation in perception tasks, thus advancing the field in collaborative approaches and information fusion.

However, most of V2V and V2I methods build V2X perception model from vehicle vision, which is insufficient in the dynamic traffic environment[73,181,182]. In this chapter, we aim to construct a road-to-vehicle vision to address the challenge of inadequate adaptability of collaborative perception models in dynamic environments.

8.2 Methodology

8.2.1 Overview

We study the V2X perception task with RSU placed on dynamic intersection scenes. In any scene, the V2X perception consists of an RSU and vehicles. The RSU and vehicles collect point cloud data, these input single-view point cloud can be converted to bird's-eye-view (BEV)[58] maps $\mathcal{V} = \{V_0, V_1, V_2, ..., V_i\}$, where V_0 for RSU and V_i for the i-th vehicle ($i > 0$). Existing V2X perception technologies struggle to effectively coordinate complex dynamic traffic data[173], which falls short of meeting the safety requirements in dynamic traffic environments: Intra-scene changes, such as pedestrians in motion and moving vehicles, along with inter-scene changes like transitions between extensive structures and road layouts across different locations, introduce disruptions to V2X perception, potentially compromising vehicle safety.

Motivated by this, this work considers building a collaborative perception model from road-to-vehicle vision for sensing complex and dynamic traffic scenarios. We name the method Adaptive Road-to-Vehicle Perception (AR2VP), as shown in Fig. 8.2, where vehicles and RSU communicate and cooperate through a broadcast communication channel. AR2VP considers addressing two kinds of scene changes:

8.2.1.1 Intra-scene changes

We first design DPR module, utilizing RSU geographical and perceptual advantages to effectively integrate the perception from vehicles, enabling vehicles to capture a more comprehensive range of dynamic factors within the scene. Then, to further enhance vehicles perception capabilities in dynamic environments, we draw inspiration from residual techniques and propose R2VPC module. Leveraging RSU perceptual advantages, this module compensates the post-collaborative perception of vehicles, filling in intra-scene dynamic factors that are overlooked by the vehicles, thereby further enhancing the overall adaptability of vehicles to dynamic environments. Lastly, it enables vehicles to adapt to large-scale scene transitions beyond intra-scene changes.

8.2.1.2 Inter-scene changes

We introduce RSU-Prompt Replay, leveraging the perceptual completeness of RSU to train learnable scene prompts for summarizing information about the entire intersection scene. This allows RSU to extract a small number of representative samples that characterize the intersection, enabling effective suppression of scene distribution drift with limited RSU storage capacity. This adaptation ensures that AR2VP can robustly perceive vehicles while accommodating variations between scenes.

8.2.2 Overcoming Intra-Scene Changes

8.2.2.1 Dynamic Perception Representation

Vehicle perception varies with the changes of dynamic entities within the intra-scene. To effectively coordinate these dynamic factors, this chapter proposes DPR module, which constructs a directed collaborative graph $\mathcal{G} = \{\mathcal{M}, \xi\}$ that leverages the advantages of RSU to adapt to intra-scene changes [Fig. 8.2(a)], where $\mathcal{M} = \Phi_{shared}(\mathcal{V})$ is encoded by the shared encoder $\Phi_{shared}(\cdot)$ to generate the feature maps and $\mathcal{V}$ represents BEV maps. The collaborative graph has three stages for dynamic perception representation.

138　Continual Artificial Intelligence towards Changing Environment

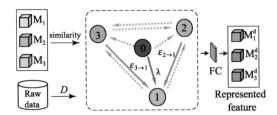

(a) Dynamic Perception Representation. For the feature map M_1, we calculate the weights ξ by combining feature similarity with the RSU-Vehicle distances $\mathcal{D}=\{d_1, d_2,...,d_i\}$ to construct a directed collaborative graph $\{\mathcal{M},\xi\}$, followed by perception representing of the feature map M_i^d.

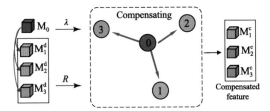

(b) Road-to-Vehicle Perception Compensating. For the feature map M_1^d, utilizing RSU-vehicles similarity ratio $\mathcal{R}$ and RSU perception threshold λ to determine the extent of RSU perception compensation within M_1^d, thus obtaining the compensated feature map M_1^c.

Fig. 8.2　Dynamic perception representation and Road-to-Vehicle perception compensating

Stage S1: position information transforming. In this stage, each vehicle transfers the position to the RSU for interaction. RSU and each vehicle has its own independent position $\mathcal{P}=\{(x_0,y_0),(x_1,y_1),(x_2,y_2),...,(x_n,y_n)\}$. When selecting the position of the i-th vehicle for collaboration, we need to transform the position information of RSU (x_0,y_0) into $(x_{0\to i}, y_{0\to i})$ corresponding to the i-th vehicles using the position matrix:

$$x_{0\to i}=\mathbf{R}_i\mathbf{R}_0^T(x_0-x_i), y_{0\to i}=\mathbf{R}_i\mathbf{R}_0^T(y_0-y_i) \quad (8.1)$$

where $\mathbf{R}_i$ denotes the rotation matrix of the i-th vehicle, which represents the orientation of the coordinate system relative to the reference coordinate system. Note that $\mathbf{R}_0$ is RSU's rotation matrix. Then, we pass the converted position information to the S2 stage to obtain the edge weights.

Stage S2: position-guided feature fusing. In this stage, each vehicle receives effective perception information from both RSU and other vehicles in the same scene. RSU, due to its unique geographical position, offers vehicles dynamic environmental adaptability. Therefore, based on the position information from Stage S1, we combine the relative distance between RSU and vehicles with the feature information between vehicles and carry out effective collaborative perception. In the directed collaborative graph $\mathcal{G}$, to determine edge weights ξ, we firstly obtain the distances $\mathcal{D} = \{d_1, d_2, \ldots, d_i\}$ between vehicles and RSU from the Stage S1:

$$d_i = \sqrt{(x_{0 \to i} - x_i)^2 + (y_{0 \to i} - y_i)^2}. \tag{8.2}$$

Then, we associate the features of different vehicles. In other words, the matrix value of the edge weight from the 2nd vehicle to the 1st vehicle $\xi_{2 \to 1}$:

$$\xi_{2 \to 1} = \frac{d_2 \cdot \cos(M_1, M_2)}{\sum_{i=2}^{N} d_i \cdot \cos(M_1, M_i)} \tag{8.3}$$

where norm($\cdot$) represents set normalization, and cos($\cdot$) represents feature similarity.

For the edge weights between RSU and vehicles, we use a fixed weight $\lambda = \frac{1}{N}$ to retain the stable perception information of RSU, and N represents the number of agents. The theoretical foundation for pre-setting λ is inspired by multi-agent coordination techniques[183], where key agents act as coordinators. In this chapter, RSU plays a role highly similar to that of key agents, serving as coordinators in the information collaboration process. Thus, we use $\frac{1}{N}$ as a coordination parameter, aiming to provide certain fairness in collaboration weight. Through Stage S2, we complete the construction of graph $\mathcal{G}$.

Stage S3: feature information aggregating. In this stage, we utilize the directed collaborative graph constructed in Stage S2 to synergistically enhance the representation of each vehicle. The perception information of each vehicle and RSU is integrated to better capture dynamic entities, achieving a comprehensive perception of the entire environment. Specifically, each vehicle aggregates the normalized edge-weighted features of all other vehicles. The updated feature map of the i-th vehicle is $\hat{\boldsymbol{M}}_i$:

$$\hat{\boldsymbol{M}}_i = \sum_{j=1}^{N} \xi_{j \to i} \boldsymbol{M}_i + \lambda \boldsymbol{M}_0. \tag{8.4}$$

In the DPR module, this study leverages the perceptual and geographical advantages of RSU to assist vehicles in perception fusion. This approach enables the perception model to initially adapt to dynamic environments, achieving a comprehensive perception effect.

8.2.2.2 Road-to-Vehicle Perception Compensating

Due to the continuous changes of scenes, using only the collaborative graph for vehicle perception in dynamic scenarios is insufficient. This chapter further leverages the advantages of RSU perceptual stability and extensive coverage to compensate for the updated vehicles perception, thereby enhancing vehicles perception in dynamic scenes. At this stage, our objective is to utilize the perception features of RSU to compensate for the updated feature maps of vehicles during the decoding process [Fig. 8.2(b)].

First, we flat the feature maps of RSU and vehicles:

$$\boldsymbol{F}^i = \text{flatten}(\boldsymbol{M}_i^d) \tag{8.5}$$

where $\boldsymbol{M}_i^d$ is obtained by decoded from $\hat{\boldsymbol{M}}_i$ using a fully-connected layer. Then, to compute the unique linear similarity between different vehicles and RSU $\mathcal{R} = \{r_1, r_2, ..., r_i\}$, considering the partial overlap in the perspectives of different vehicles, we employed the Pearson correlation coefficient calculation method[184]. Before the calculation, we centralized the data to eliminate the translational effect of redundant data, thereby emphasizing the unique correlation between features more prominently, which is to accurately determine which feature map needs to be compensated by using RSU perception, avoiding unnecessary computational burdens. The formula is as follows:

$$r_i = \frac{\sum\limits_{i=1}^{n} a_i}{\sqrt{\sum\limits_{i=1}^{n} a_0^2 \cdot a_i^2}} \qquad (8.6)$$

where

$$a_i = \sum_{j=1}^{n}\left(\boldsymbol{F}_j^i - \sum_{i=1}^{k}\frac{\boldsymbol{M}_i^d}{k}\right) \qquad (8.7)$$

Then, a predefined threshold λ is used to determine whether compensation with RSU is required for feature mappings. We supplement them with RSU perception information, resulting in compensated feature maps $\mathcal{M}^c$:

$$\boldsymbol{M}_i^c = (\lambda - r_i)\boldsymbol{M}_0^d + \boldsymbol{M}_i^d \qquad (8.8)$$

Finally, the model is followed by different output heads based on the tasks [such as segmentation (seg.) and detection (det.)] to generate perception results. Though AR2VP adapts to the changes of intra-scene, adapting to the changes of inter-scene is still unsolved.

8.2.3 Overcoming Inter-Scene Changes

Due to the relatively abrupt changes of inter-scene, models often face the challenge of distributional adaptation drift when learning new environments[129,185]. In this situation, the model suffers from serious catastrophic forgetting[172,186], which may lead to the risk of traffic accidents. Inspired by the experience replay in continual learning, we propose the RPR method for V2X perception. RPR takes advantage of the unique perceptual completeness and stability of RSU in the scene. RPR effectively suppresses distribution adaptation drift and reduces catastrophic forgetting, but it also optimizes the storage requirements that replay methods entail.

In the RPR module, we seek to utilize limited storage of RSU to reduce the drift as much as possible. Firstly, we build prompts to extract representative frames in the V2X perception, by encapsulating the perceptual information of the entire global scene. Specifically, we first randomly initialize k prompts $\mathscr{P} = \{ \boldsymbol{P}^1, \cdots, \boldsymbol{P}^k \}$. Then, we update prompts using the Exponential Moving Average (EMA) algorithm[187] to learn various aspects of continuous scene data and reduce the potential impact of errors from a single prompt.

$$\boldsymbol{P}_t^i = \alpha \boldsymbol{S}_t + (1-\alpha) \boldsymbol{P}_{t-1}^i \quad t = 1, \cdots, n \tag{8.9}$$

where, $\alpha = 0.5$ represents the smoothing factor. To make the RSU feature computed with the $\mathscr{P}$, we pass $\boldsymbol{M}_0$ through the fully connected layer f_n to get $\boldsymbol{S}_t$,

$$\boldsymbol{S}_t = f_n(\boldsymbol{M}_0), \tag{8.10}$$

which signifies the features coalesced by RSU in the t-th frame of the scene. Employing the EMA algorithm helps decrease the accumulation of redundant information within prompts, providing a more comprehensive representation of overall scene information.

Subsequently, while learning in new scenes, based on these prompts, typical scene samples $\mathscr{S}_p$ are sliced and stored in RSU:

$$\mathscr{S}_p = \mathscr{S}_c[\text{top}_l(\cos_t(\boldsymbol{P}_t, \boldsymbol{S}_t))] \quad t = 1, 2, 3, \ldots \tag{8.11}$$

where, $\mathscr{S}_c$ represents new scene data. $\text{top}_l(\cdot)$ represents extracting the indices of the top l samples with higher global information. Therefore, different typical samples can be selected based on the similarity of the prompts.

Finally, when the model learns a new scenario again, $\mathscr{S}_p$ and model will be updated, and effective knowledge review will be achieved by revisiting these stored representative samples. This method can not only effectively alleviate scene distribution drift and reduce model forgetting, but also optimize RSU storage capacity to achieve minimum storage and robust forgetting suppression:

$$\theta = \text{SGD}(\mathscr{S}_c \cup \mathscr{S}_p, \theta). \tag{8.12}$$

This formula represents model update, where $\text{SGD}(\cdot)$ is the stochastic gradient descent algorithm, and θ represents the model parameters. And the next formula represents $\mathscr{S}_p$ update:

$$\mathscr{S}_p = \mathscr{S}_c[\text{top}_l(\cdot)] \cup \mathscr{S}_p. \tag{8.13}$$

In this module, the model engages in comprehensive learning of the previous typical data stored in RSU and the current scene data. This not only acquires knowledge from new scenes but also revisits knowledge from previous scenes, achieving a synergistic combination of learning and review. Our Prompt-replay effectively mitigates the catastrophic forgetting caused by inter-scene changes.

8.2.4 The Whole Algorithm

In the process of model learning update to dataset $\mathscr{S}$, we use L_{det} loss for the detection task to update learning:

$$L_{det} = \sum_{i=1}^{n} \frac{\eta(Y_i - Y'_i)^2}{\sigma^2} \tag{8.14}$$

where η typically takes a value of 0.5 and 0.7.

In the segmentation task, we use L_{seg} loss for update learning:

$$L_{seg} = -\sum_{i=1}^{n} (Y_i \cdot \log(Y'_i)) \tag{8.15}$$

where Y and Y' represent the label and prediction in scene $\mathscr{S}$, σ is a hyperparameter.

The overall updated loss L of the model is as follows:

$$L = L_{det/seg}^{previous} + L_{det/seg}^{current} \tag{8.16}$$

where $L_{det/seg}^{previous}$ represents the loss from the replayed data in previous scenes for the current task (det./seg.), and $L_{det/seg}^{current}$ represents the loss from current scene data used for the current task.

In the whole AR2VP research (Fig. 8.2 and Algorithm 9), we design the DPR module [Fig. 8.2(a) and Algorithm 7], merging geographical and feature data from RSU and vehicles to create an adaptable collaborative graph for dynamic scenarios. This effectively integrates perception information from different vehicles, enabling a more comprehensive grasp of dynamic elements within the scene. Subsequently, inspired by residual techniques, we propose the R2VPC module [Fig. 8.2(b) and Algorithm 7]. By leveraging RSU perceptual advantages, this module compensates post-collaborative vehicle perception, filling in intra-scene dynamic elements overlooked by the vehicles, further enhancing overall adaptability to dynamic settings. Lastly, to extend adaptability beyond intra-scene changes, we introduce Prompt-replay (Fig. 8.3 and Algorithm 8), in conjunction with RSU perceptual completeness and continual learning techniques, to alleviate the issue of distribution adaptation drift within the model. This empowers AR2VP to cope with inter-scene changes, ensuring robust and reliable vehicle perception. In general, in the context of intra-scene dynamic perception learning, we employ the full-scenario training method, exclusively utilizing Algorithm 7. When dealing with inter-scene transformation perception learning, we adopt the cross-scene learning training approach, incorporating both Algorithm 7 and Algorithm 8.

Algorithm 7: Intra-scene adaptation
Input: BEV maps $\mathcal{V}$, position $\mathcal{P}$
Output: Compensated feature maps $\mathcal{M}^c$
1   Encoder($\mathcal{V}$)→$\boldsymbol{M}$ // Agent features;
2   $\mathcal{P}$→Eq.(8.2)→$\mathcal{D}$ // Vehicle-RSU distances;
3   ($\boldsymbol{M},\mathcal{D}$)→Eq.(8.3)→$\xi_{i\to j}$ // Correlations between agents;
4   ($\xi_{i\to j},\boldsymbol{M}$)→Eq.(8.4)→$\hat{\mathcal{M}}$ // Update features;
5   $\hat{\mathcal{M}}$→Eq.(8.5) and (8.6)→$\mathcal{R}$ // Feature similarity ratio;
6   ($\mathcal{R},\hat{\mathcal{M}}$)→Eq.(8.8)→$\boldsymbol{M}^c$ // Improve vehicle perception with RSU;
7   return $\mathcal{M}^c$

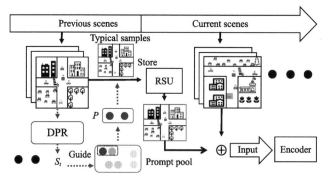

After calculation by the DPR model, the feature information $\boldsymbol{S}_t$ of RSU of each frame is obtained, and then $\boldsymbol{S}_t$ is used to guide the training of prompt $\mathcal{P}$ to participate in the extraction of typical samples in the previous scenes. Finally, the Current scenes and typical samples are input into the encoder for model training.

Fig. 8.3 Prompt-replay

Dynamic V2X Perception from Road-to-Vehicle Vision

Algorithm 8: Inter-scene adaption

Input: RSU feature $\boldsymbol{S}_t$, new scene data $\mathscr{S}_c$
Output: A few typical scene data $\mathscr{S}_p$

1   if The first scene then
2         $(\boldsymbol{S}_t, \mathscr{S}_c) \to$ Eq.(8.9) $\to \boldsymbol{P}$ // Update prompts;
3         $(\boldsymbol{P}, \boldsymbol{S}_t) \to$ Eq.(8.11) $\to$ **sim**   // Calculate feature similarity;
4         $\mathscr{S}_c \to$ Eq.(8.11) $\to \mathscr{S}_p$ // Extract typical samples;
5   end
6   else
7         $(\mathscr{S}_c, \mathscr{S}_p) \to$ Eq.(8.13) $\to \mathscr{S}_p$ // Update typical samples;
8   end
9   return $\mathscr{S}_p$;

Algorithm 9: AR2VP

Input: New scene data $\mathscr{S}_c \subset (\mathscr{V}, \mathscr{P})$
Output: Prediction Y'_i

1   for each scene $i \in [0, n]$ do
2   if $i = 0$ then
3            $\mathscr{S}_c \to$ Alg.7 $\to \mathscr{M}^c$;   // Get compensated feature maps
4   Alg.8 $\to \mathscr{S}_p \Rightarrow$ RSU;   // Store typical samples
5   end
6   else
7   RSU $\Rightarrow \mathscr{S}_p$;   // Get typical samples
8            $(\mathscr{S}_p, \mathscr{S}_c) \to$ Alg.7 $\to \mathscr{M}^c$;
9   Alg.8 $\to \mathscr{S}_p \Rightarrow$ RSU;   // Update and store typical samples
10   end
11      $\mathscr{M}^c \to Y'_i$ // Get prediction
12   end
13   return Y'_i;

8.2.5 Bandwidth Discussion

We speculate and analyze the vehicle-road cooperative communication in two scenarios based on the presence or absence of additional computational power in RSU, as shown in Fig. 8.4 and Fig 8.5.

If RSU has normal computational capabilities (containing only encoders), vehicles communicate with each other. Communication details are outlined in Fig. 8.4.

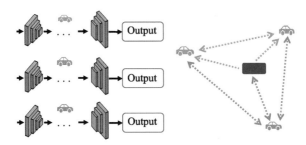

When RSU lacks additional computational power, vehicles will store the entire model. One vehicle communicates $2n$ times, and the entire collaborative system requires communication in the order of $2n^2$ times.

Fig. 8.4 V2X communication for the RSU without additional computational power

If RSU has extra computational power (including decoders), we store the entire model in the RSU. This exchange of additional RSU computational power is used to gain advantages in vehicle-road cooperative communication, as detailed in Fig. 8.5. Therefore, regarding the two scenarios mentioned above, we can draw conclusions. Storing the model in the RSU further reduces n^2 communication bandwidth.

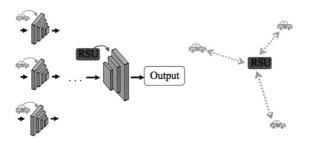

When RSU has additional computational power, vehicles will only store the encoder model. This will significantly reduce communication between vehicles. One vehicle communicates 2 times, and the entire collaborative system requires communication in the order of $2n$ times.

Fig. 8.5 V2X communication for the RSU with additional computational power

AR2VP can be optimized not only by leveraging the additional computational power of RSU to increase or reduce communication bandwidth but also by optimizing data transmission to save the communication bandwidth. Since AR2VP employs an intermediate fusion approach, the data transmitted in terms of communication consists of feature information, where RSU and vehicles could compress their feature map prior to transmission. Optionally, in our study, we make use of a 1×1 convolutional autoencoder[189] to compress and decompress the feature maps along the channel dimension. The autoencoder is trained together with the whole system. Therefore, AR2VP can increase communication efficiency by a factor of n through feature compression, where n represents the degree of feature compression. We show this in our experiments (see the subsection 8.3.6).

8.3 Experiments

8.3.1 Data Preparation

In this study, we evaluated V2X perception tasks offline using the V2X-sim dataset. The V2X-sim dataset emulates multi-agent scenarios, wherein each scenario encompasses a 20-second traffic flow across multiple intersections. Laser radar recordings are captured at intervals of 0.2 seconds, yielding a total of 100 frames per scenario. This dataset comprises 100 distinct scenes, with each frame housing multiple samples. The training set comprises 23,500 samples, while the test set contains 3,100 samples. To establish a fixed large scenario, we selected 30 scenes, which collectively contribute 3,000 frames. Among these, the training set comprises 2,700 frames, while the test set consists of 300 frames. Moreover, to implement cross-scene experiments, we also train V2X model sequentially on three major scenes in chronological order. Fig. 8.6(a) shows the details of the dataset and Fig. 8.6(b) shows three major scenes.

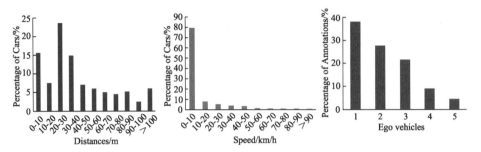

(a) Left: Statistics of the distance between every two ego vehicles for all frames. Middle: Speed of cars located within 70m from ego vehicles. Right: Percentage of annotated vehicles observed by 1−5 ego vehicles.

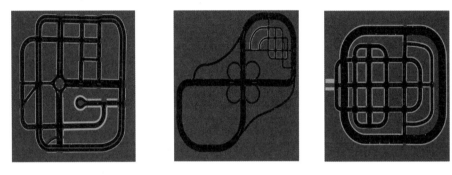

(b) The map of towns which are squared-grid towns with cross junctions and multiple lanes per direction[188].

Fig. 8.6   Dataset illustration

8.3.2 Evaluation Metric

In this chapter, we evaluate our method on two V2X perception tasks, including scene segmentation and vehicle objection. We employ the generic BEV detection evaluation metric, Average Precision (AP)[187] at Intersection-over-Union (IoU) threshold of 0.5 and 0.7:

$$AP=(TP/(TP+FP+FN)) \qquad (8.17)$$

where True Positives (TP) refers to the number of correctly detected objects by the model. This is determined by comparing the predicted bounding boxes with the ground truth bounding boxes in terms of their overlap. If the predicted bounding box has sufficient overlap with any of the ground truth bounding boxes, it is considered a true positive. False

Positives (FP) refers to the number of objects incorrectly detected by the model, such as identifying background or non-existing objects as targets. This is determined by examining the overlap between predicted bounding boxes and all ground truth bounding boxes. If the overlap between the predicted bounding box and all ground truth bounding boxes is insufficient, it is considered a false positive. False Negatives (FN) refers to the number of real objects missed by the model. If the ground truth bounding box of an object has no overlap with any predicted bounding box or has insufficient overlap, it is considered a false negative.

We evaluate the segmentation performance using meanIoU (mIoU). We evaluate the extent of forgetting across inter-scene changes using Forget. IoU threshold of 0.5 and 0.7:

$$\mathrm{mIoU} = \sum_{i=1}^{N}(\mathrm{IoU}_i)/N \qquad (8.18)$$

where IoU_i is the IoU for each agent. N is the total number of agents.

8.3.3 Compared Methods

We first compare with the early fusion method[190] and the late fusion method. Furthermore, four intermediate fusion methods are used, including When2com[191] which uses asymmetric attention for image segmentation communication, Who2com[192], V2V[166] which employs spatial-aware graph neural networks for autonomous driving perception, and Disco[167] which offers adaptive perception with a matrix-weighted collaboration graph. Since the original Who2com and When2com do not consider pose information, we consider both pose-aware and pose-agnostic versions (with *) to achieve fair comparisons. All the methods use the same segmentation and detection backbones and conduct collaboration at the same intermediate feature layer.

We utilize data stored in RSU (MB) to mitigate the forgetting issue caused by inter-scene variations. These stored data consist of two types: firstly, the original old scene data from RSU-Random Replay (RRR) methods, and secondly, the prompt data proposed in RSU-Prompt Replay (RPR). Leveraging these lightweight Prompt data can to some extent substitute the old scene data, thereby reducing storage capacity and computational burden.

8.3.4 Comparisons under Intra-Scene Changes

Tables 8.1 and 8.2 show the comparisons in terms of mIoU (seg.) and AP@0.5/0.7 (det.). Compared to the pose-aware When2com, AR2VP improves by 57.47% in segmentation of unlabeled data and 25.30% in mIoU. Compared to Disco, AR2VP improves by 4.64% in mIoU. Compared to the When2com, AR2VP improves by 13.15% in AP@0.5 and 12.75% in AP@0.7. Compared to Disco, AR2VP improves by 2.49% in AP@0.5 and 2.36% in AP@0.7. The qualitative results are shown in Fig. 8.7(a) (seg.) and Fig. 8.7(b) (det.).

Table 8.1 Segmentation comparison of intra-scene

Unit: %

Method	Unlabeled	Vehicles	Sidewalk	Ground	Road	Buildings	Pedestrian	Vegetation	mIoU
Early Fusion	65.96	90.87	94.67	94.52	97.37	94.89	50.45	90.26	84.87
Late Fusion	48.56	72.17	86.88	85.96	93.48	85.92	18.21	80.07	71.41
When2com	41.25	65.47	69.62	58.83	83.65	62.36	27.18	62.00	58.79
When2com*	41.42	63.47	72.19	58.81	81.02	68.55	28.18	74.36	59.75
Who2com	42.25	66.47	70.62	59.83	84.65	63.36	28.18	63.00	59.80
Who2com*	40.02	63.47	72.60	62.81	81.00	60.55	28.20	66.36	60.75
V2V	60.10	84.92	93.04	91.87	95.98	93.10	33.89	86.85	79.01
Disco	61.15	84.75	92.82	92.62	96.52	92.95	35.49	87.01	80.41
AR2VP	98.89	85.31	93.37	92.86	96.63	93.31	33.63	86.38	85.05

The best options in intermediate collaboration methods are shown in bold and * indicates the pose-aware version.

Table 8.2 Detection performance in comparison with other methods

Unit: %

Method	AP	
	@IoU=0.5	@IoU=0.7
Early Fusion	96.63	96.05
Late Fusion	85.62	83.84

continued

Method	AP	
	@IoU=0.5	@IoU=0.7
When2com	81.35	80.02
When2com*	81.86	80.69
Who2com	81.32	79.98
Who2com*	81.69	80.66
V2V	91.89	89.90
Disco	92.01	90.41
AR2VP	94.50	92.77

* indicates the pose-aware version.

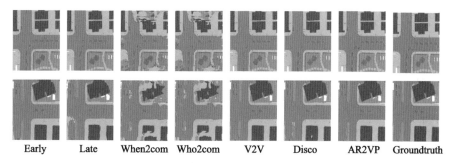

(a) Visualizations of collaborative BEV semantic segmentation.

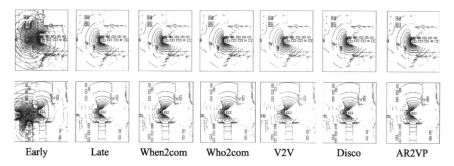

(b) Visualizations of BEV detection on V2X-Sim (Red and green boxes are the predictionsand ground-truths respectively).

Fig. 8.7 Visualizations of collaborative BEV

We observed that AR2VP demonstrates superior entity perception outcomes, achieving the highest overall perception performance. This analysis underscores that current V2X technologies rarely rely on RSU to expand perception horizons. In contrast, AR2VP harnesses the latent strengths of RSU to address intra-scene changes, which enhances the vehicle's ability to adapt to dynamic scenes, consequently elevating the overall perception capabilities. However, AR2VP does exhibit a performance drawback in pedestrian detection, implying a particular challenge in detecting small targets.

8.3.5 Comparisons under Inter-Scene Changes

Table 8.3 shows the comparison on inhibition of forgetting across different scenes. Compared to the pose-aware When2com, AR2VP improves by 28.35% in mIoU and reduces the forgetting rate by 18.55%. Compared to Disco, AR2VP improves by 26.04% in mIoU and reduces the forgetting rate by 33.23%. Compared to the pose-aware When2com, AR2VP improves by 29.34% in AP@0.5 and 28.58% in AP@0.7, and reduces forgetting rate by 26.27% in AP@0.5 and 25.09% in AP@0.7. Compared to Disco, AR2VP improves by 7.51% in AP@0.5 and 8.10% in AP@0.7 and reduces the forgetting rate by 9.28% in AP@0.5 and 10.86% in AP@0.7. Furthermore, AR2VP, when utilizing Prompt-replay, maintains robust perceptual capabilities even when subjected to minimal usage of 110MB, outperforming RSU-Random Replay (RRR) in similar conditions.

Table 8.3 Perception comparisons on inter-scene changes

Method	Det.(@IoU=0.5)		Det.(@IoU=0.7)		Seg.		Mem.(MB)↓
	AP/%↑	Forget/%↓	AP/%↑	Forget/%↓	mIoU/%↑	Forget/%↓	
Early Fusion	85.15	14.96	81.03	19.19	52.71	42.62	0
Late Fusion	62.77	27.61	57.86	31.96	40.14	31.91	0
When2com	61.40	32.23	57.90	34.44	40.73	30.76	0
When2com*	61.64	31.38	57.98	33.01	45.60	28.65	0
Who2com	61.04	32.89	57.66	34.63	41.63	31.56	0
Who2com*	61.47	32.01	58.11	33.79	44.71	29.56	0
V2V	82.54	15.65	76.98	20.63	48.50	46.08	0

continued

Method	Det.(@IoU=0.5)		Det.(@IoU=0.7)		Seg.		Mem.(MB)↓
	AP/%↑	Forget/%↓	AP/%↑	Forget/%↓	mIoU/%↑	Forget/%↓	
Disco	83.47	14.39	78.46	18.78	48.09	43.31	0
V2V(RRR)	87.63	7.38	83.48	9.46	63.06	18.09	240
Disco(RRR)	87.65	7.61	83.65	9.53	64.74	19.12	240
AR2VP(RRR)	89.52	6.16	84.90	8.84	71.51	11.48	240
AR2VP(RPR)	90.98	5.11	86.56	7.92	74.13	10.10	240
AR2VP(RPR)*	90.23	5.71	85.78	8.45	73.66	10.54	130

RSU-Random Replay (RRR) represents the random extraction of old scene samples, RSU-Prompt Replay (RPR) represents the extraction of typical samples using prompts. * indicates the pose-aware version.

AR2VP presents itself as a frontrunner in terms of overall perception performance. Upon analysis, it's evident that traditional V2X technologies disregard the influence of inter-scene changes on perception. In contrast, AR2VP exploits the storage capacity of RSU and integrates continual learning principles to effectively address inter-scene changes. This strategic approach empowers vehicles to assimilate new scenes while minimizing the extent of memory loss from prior scenes. This capability shows a strong adaptability to inter-scene changes in perception, thereby enhancing the global robustness of perception.

The prompt integrates RSU features across different time steps to represent dynamic panoramic information, thereby enhancing the existing perception system's resistance to forgetting. Compared to other methods, APVP focuses more on utilizing RSU perception data for dynamic scene understanding. Therefore, although Prompt-replay can be applied to other models to mitigate forgetting, AR2VP notably demonstrates the most favorable suitability. Fig. 8.8 shows the learning processes of new scene data based on the old model, revealing the degree of memory forgetting from previous scenes. The observation is clear: V2V and Disco struggle to accommodate inter-scene changes, leading to significant memory loss from previous scenes. In contrast, AR2VP adeptly navigates inter-scene changes, exhibiting a higher retention of memories from prior scenes. This analysis underscores AR2VP's capacity for the lowest forgetting rate and the most proficient performance in addressing inter-scene changes.

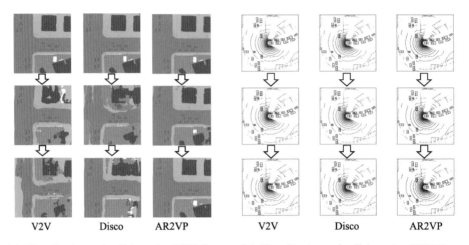

(a) Visualizations of collaborative BEV Seg. on inhibition of forgetting.

(b) Visualizations of collaborative BEV Det. on inhibition of forgetting.

Fig. 8.8　Visualizations of collaborative BEV on inhibition of forgetting

8.3.6　Performance-Bandwidth Trade-off Analysis

In Fig. 8.9, we compare the proposed AR2VP with the baseline methods in terms of the trade-off between segmentation performance and communication bandwidth. The dashed line represents the baseline based on when2com. To show the better trade-off of the proposed AR2VP, we employ an autoencoder to compress features and reduce the communication bandwidth used for feature transmission (AR2VP/x means compress x times). Compared to AR2VP, AR2VP/32 degrades by 1.31% in segmentation, still outperforming Disco in terms of mIoU. This means the compressed features in AR2VP do not significantly compromise perception performance.

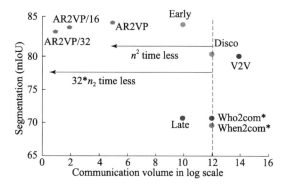

RSU demonstrates stability in perception and possesses geographical advantages in inter-scene changes, as opposed to the continual mobility of vehicles in intra-scene dynamics. With storage and communication capabilities, RSU stores old scenes data and model, enabling adaptation to inter-scene changes.

Fig. 8.9 Dynamic scene advantage of RSU

8.3.7 Ablation Study

We conduct ablation studies to analyze the perceptual performance of the graph and the communication in the presence and absence of RSU. The results are shown in Table 8.4. Firstly, we find that the participation of RSU in the collaborative process provides additional perception coverage to enhance vehicle perception performance. Secondly, the collaborative graph effectively integrates all perception information, enabling vehicles to comprehensively perceive entities within the scene. Finally, in scenarios where RSU is present, the compensator utilizes the stable perception information from RSU to effectively compensate for vehicle perception. Moreover, the compensator benefits from the presence of RSU, showing a positive effect. In the absence of RSU, using vehicles to compensate for other vehicles' perceptions would lead to negative consequences.

Table 8.4 Ablation studies of intra-scene

RSU	graph	compensator	Det./%		Seg./%
			AP@0.5	AP@0.7	mIoU
✗	✗	✓	66.99	65.47	55.01
✗	✓	✗	89.88	87.95	75.65

continued

RSU	graph compensator		Det./%		Seg./%
			AP@0.5	AP@0.7	mIoU
✗	✓	✓	90.65	88.46	73.06
✓	✗	✓	68.56	66.32	56.47
✓	✓	✗	93.80	91.71	84.04
✓	✓	✓	94.50	92.77	85.05

Also, we conduct ablation studies on the initial quantity of prompts and the storage capacity of RSU. The results are shown in Table 8.5. Our findings indicate that: (1) An inappropriate quantity of prompts makes it challenging to adequately capture the features of the entire scene, leading to the extraction of representative samples that are ineffective in suppressing inter-scene distribution drift issues. (2) Theoretically, the model's robustness is improved with an increase in RSU storage capacity.

Table 8.5 Ablation studies of inter-scene

Prompt	Seg./%		
	80(MB)	130(MB)	240(MB)
2	67.98	70.86	72.42
3	68.71	73.66	74.13
4	68.58	73.53	74.62

8.4 Chapter Conclusion

In this chapter, we propose a vehicle-road cooperative perception model, named AR2VP, which is capable of adapting to dynamic environments. It mainly consists of a DPR module, an R2VPC module, and an RPR method. The DPR module efficiently integrates vehicle perceptions to comprehensively capture dynamic factors within the scene, enhancing the perception capabilities of the collaborative perception model. The R2VPC module is geared towards effectively retaining the RSU perception information, especially in the face of intra-scene changes. The RPR method integrates within the RSU's storage capacity and facilitates the retention of a small volume of historical scene data. This approach ensures that the cooperative model maintains a certain level of robustness when confronted with inter-scene changes. Comprehensive experiments demonstrate that AR2VP achieves adaptability to dynamic environments and an appealing performance-bandwidth tradeoff through a more direct design principle.

As for limitation and future work, in our experiments, we find that AR2VP is not very robust to small objects, which may be failed recognition or forgotten. We believe the issue is raised by the perception conflicts from different vehicles. In the future, based on our experimental findings, we intend to enhance the AR2VP's capability to recognize small objects. This will involve further refinement of the model to ensure more accurate and effective identification of small entities.

Chapter 9

Conclusions and Future Work

In this book, we present the theory and applications of continual artificial intelligence, as well as some of our research work in this field. Continual learning enables a machine learning system to adaptively operate in a changing environment. It is an important research direction in artificial intelligence in recent years and one of the core technologies for general artificial intelligence. In this book, we mainly summarize the scenarios, difficulties, solutions of continual learning, and some applications of continual learning in multilabel classification problems, edge computation scenarios, and few-shot classification. The study of continual learning will enable artificial intelligence to handle changing data at a small cost in various scenarios.

We propose a cross-task margin loss function, which connects each isolation task and sets two-level angular margins in the cross-entropy to encourage the intra-class/task compactness and the inter-class/task discrepancy. We also propose a centroid-based rehearsal selection strategy to sample more representative samples in rehearsal-based continual learning. In addition, we propose to evaluate the example influence in continual learning and to use the example influence to improve the training. We also propose an asymmetric gradient discrepancy metric for parallel continual learning, and an augmented graph neural network for multi-label continual learning. We propose a federated continual learning based on Fisher information fusion. Finally, we propose to apply continual learning to V2X perception application.

The study of continual learning is still in its infancy, and there are many problems that need to be solved. In the future, we will continue to study the theory and application of continual learning and explore more effective solutions to the problems of catastrophic forgetting, sample inefficiency, and task interference. We will also study the application of continual learning in more fields, such as natural language processing, robotics, and intelligent transportation.

References

[1] R. Girshick. "Fast r-cnn," in Proceedings of the IEEE International Conference on Computer Vision, 2015: 1440—1448.

[2] K. He, G. Gkioxari, P. Dollár, et al. "Mask r-cnn," in Proceedings of the IEEE International Conference on Computer Vision, 2017: 2961—2969.

[3] S. Humeau, K. Shuster, M. -A. Lachaux, et al. "Poly-encoders: Architectures and pre-training strategies for fast and accurate multi-sentence scoring," arXiv preprint arXiv: 1905.01969, 2019.

[4] L. Pang, Y. Lan, J. Guo, et al. "Text matching as image recognition," in Proceedings of the AAAI Conference on Artificial Intelligence, 2016, Vol. 30, No. 1: 2793—2799.

[5] A. A. Shvets, A. Rakhlin, A. A. Kalinin, et al. "Automatic instrument segmentation in robot-assisted surgery using deep learning," in IEEE International Conference on Machine Learning and Applications, 2018: 624—628.

[6] N. Sünderhauf, O. Brock, W. Scheirer, et al. "The limits and potentials of deep learning for robotics," The International Journal of Robotics Research, 2018, Vol. 37, No. 4—5: 405—420.

[7] H. -T. Cheng, L. Koc, J. Harmsen, et al. "Wide & deep learning for recommender systems," in Proceedings of the 1st Workshop on Deep Learning for Recommender Systems, 2016: 7—10.

[8] A. Dosovitskiy, L. Beyer, A. Kolesnikov, et al. "An image is worth

16×16 words: Transformers for image recognition at scale," in International Conference on Learning Representations, 2020.

[9] R. Hu and A. Singh. "Unit: Multimodal multitask learning with a unified transformer," in Proceedings of the IEEE International Conference on Computer Vision, 2021: 1439−1449.

[10] S. B. Kotsiantis, I. Zaharakis, P. Pintelas, et al. "Supervised machine learning: A review of classification techniques," Emerging Artificial Intelligence Applications in Computer Engineering, 2007, Vol. 160, No. 1: 3−24.

[11] A. Krizhevsky, I. Sutskever, and G. E. Hinton. "Imagenet classification with deep convolutional neural networks," Communications of the ACM, 2017, Vol. 60, No. 6: 84−90.

[12] K. He, X. Zhang, S. Ren, et al. "Deep residual learning for image recognition," in Proceedings of the IEEE Conference on Computer Vision and Pattern Recognition, 2016: 770−778.

[13] A. Vaswani, N. Shazeer, N. Parmar, et al. "Attention is all you need," in Advances in Neural Information Processing Systems, 2017.

[14] G. M. van de Ven and A. S. Tolias. "Three scenarios for continual learning," in Advances in Neural Information Processing Systems Continual Learning Workshop, 2018.

[15] S. -A. Rebuffi, A. Kolesnikov, G. Sperl, et al. "iCaRL: Incremental classifier and representation learning," in Proceedings of the IEEE Conference on Computer Vision and Pattern Recognition, 2017: 2001−2010.

[16] R. Aljundi, M. Lin, B. Goujaud, et al. "Gradient based sample selection for online continual learning," in Advances in Neural Information Processing Systems, 2019.

[17] J. Bang, H. Kim, Y. Yoo, et al. "Rainbow memory: Continual learning with a memory of diverse samples," in Proceedings of the IEEE Conference on Computer Vision and Pattern Recognition, 2021: 8218−8227.

[18] J. Kirkpatrick, R. Pascanu, N. Rabinowitz, et al. "Overcoming catastrophic forgetting in neural networks," Proceedings of the National Academy of Sciences of the United States of America, 2017, Vol. 114, No. 13: 3521−3526.

[19] T. Li, A. K. Sahu, M. Zaheer, et al. "Federated optimization in heterogeneous networks," in Proceedings of Machine Learning and Systems, 2020: 429−450.

[20] L. Yang, L. Li, Z. Zhang, et al. "Dpgn: Distribution propagation graph network for few-shot learning," in Proceedings of the IEEE Conference on Computer Vision and Pattern Recognition, 2020: 13390−13399.

[21] H. Yao, Y. Wei, J. Huang, et al. "Hierarchically structured meta-learning," in International Conference on Machine Learning, 2019: 7045−7054.

[22] S.-e. Yoon, H. Song, K. Shin, et al. "How much and when do we need higher order information in hypergraphs? A case study on hyperedge prediction," in Proceedings of The Web Conference 2020, 2020: 2627−2633.

[23] Z. Li and D. Hoiem. "Learning without forgetting," in Proceedings of the European Conference on Computer Vision, 2016: 614−629.

[24] X. Gong, J. Yang, D. Yuan, et al. "Generalized large margin knn for partial label learning," IEEE Transactions on Multimedia, 2022, Vol. 24: 1055−1066.

[25] G. Yang, E. Fini, D. Xu, et al. "Continual attentive fusion for incremental learning in semantic segmentation," IEEE Transactions on Multimedia, 2023, Vol. 55: 3841−3854.

[26] A. Robins. "Catastrophic forgetting, rehearsal and pseudorehearsal," Connection Science, 1995, Vol. 7, No. 2: 123−146.

[27] D. Liu, F. Lyu, L. Li, et al. "Centroid distance distillation for effective rehearsal in continual learning," in IEEE International Conference on Acoustics, Speech and Signal Processing (ICASSP),

2023: 1−5.

[28] J. S. Smith, J. Tian, S. Halbe, et al. "A closer look at rehearsal-free continual learning," in Proceedings of the IEEE Conference on Computer Vision and Pattern Recognition, 2023: 2409−2419.

[29] F. Lyu, S. Wang, W. Feng, et al. "Multi-domain multi-task rehearsal for lifelong learning," in Proceedings of the AAAI Conference on Artificial Intelligence, 2021, Vol. 35, No. 10: 8819−8827.

[30] C. Atkinson, B. McCane, L. Szymanski, et al. "Pseudo-rehearsal: Achieving deep reinforcement learning without catastrophic forgetting," Neurocomputing, 2021, Vol. 428: 291−307.

[31] H. Shah, K. Javed, and F. Shafait. "Distillation techniques for pseudo-rehearsal based incremental learning," arXiv preprint arXiv: 1807.02799, 2018.

[32] J. Pomponi, S. Scardapane, and A. Uncini. "Pseudo-rehearsal for continual learning with normalizing flows," arXiv preprint arXiv: 2007.02443, 2020.

[33] Z. Wang, Z. Zhang, S. Ebrahimi, et al. "Dualprompt: Complementary prompting for rehearsal-free continual learning," in Proceedings of the European Conference on Computer Vision, 2022: 631−648.

[34] D. Rolnick, A. Ahuja, J. Schwarz, et al. "Experience replay for continual learning," arXiv preprint arXiv: 1811.11682, 2018.

[35] A. Chaudhry, M. Ranzato, M. Rohrbach, et al. "Efficient lifelong learning with a-gem," in Proceedings of the International Conference on Learning Representations, 2019.

[36] Z. Ye, F. Hu, F. Lyu, et al. "Disentangling semantic-to-visual confusion for zero-shot learning," IEEE Transactions on Multimedia, 2021, Vol. 24: 2828−2840.

[37] R. Aljundi, P. Chakravarty, and T. Tuytelaars. "Expert gate: Lifelong learning with a network of experts," in Proceedings of the

IEEE Conference on Computer Vision and Pattern Recognition, 2017: 3366—3375.

[38] A. Mallya and S. Lazebnik. "Packnet: Adding multiple tasks to a single network by iterative pruning," in Proceedings of the IEEE Conference on Computer Vision and Pattern Recognition, 2018: 7765—7773.

[39] A. Mallya, D. Davis, and S. Lazebnik. "Piggyback: Adapting a single network to multiple tasks by learning to mask weights," in Proceedings of the European Conference on Computer Vision, 2018: 67—82.

[40] J. Serra, D. Suris, M. Miron, et al. "Overcoming catastrophic forgetting with hard attention to the task," in Proceedings of the International Conference on Machine Learning, 2018: 4548—4557.

[41] R. M. French. "Catastrophic forgetting in connectionist networks," Trends in Cognitive Sciences, 1999, Vol. 3, No. 4: 128—135.

[42] M. De Lange, R. Aljundi, M. Masana, et al. "Continual learning: A comparative study on how to defy forgetting in classification tasks," arXiv preprint arXiv: 1909.08383, 2019.

[43] D. Lopez-Paz and M. Ranzato. "Gradient episodic memory for continual learning," in Advances in Neural Information Processing Systems, 2017: 6467—6476.

[44] P. Dhar, R. V. Singh, K. -C. Peng, et al. "Learning without memorizing," in Proceedings of the IEEE Conference on Computer Vision and Pattern Recognition, 2019: 5138—5146.

[45] S. Thrun. "Lifelong learning algorithms," Learning to learn. 1998, Springer: 181—209.

[46] Y. Guo, M. Liu, T. Yang, et al. "Learning with long-term remembering: Following the lead of mixed stochastic gradient," arXiv preprint arXiv: 1909.11763, 2019.

[47] Y. Yang and T. M. Hospedales. "A unified perspective on multi-domain and multi-task learning," in International Conference on

Learning Representations, 2014.

[48] Q. Sun, F. Lyu, F. Shang, et al. "Exploring example influence in continual learning," in Advances in Neural Information Processing Systems, 2022.

[49] S. Tang, D. Chen, L. Bai, et al. "Mutualcrf-gnn for fewshot learning," in Proceedings of the IEEE Conference on Computer Vision and Pattern Recognition, 2021: 2329−2339.

[50] A. Chaudhry, A. Gordo, P. Dokania, et al. "Using hindsight to anchor past knowledge in continual learning," in Proceedings of the AAAI Conference on Artificial Intelligence, 2021, Vol. 35, No. 8: 6993−7001.

[51] X. Lin, H. -L. Zhen, Z. Li, et al. "Pareto multi-task learning," in Advances in Neural Information Processing Systems, 2019.

[52] O. Sener and V. Koltun. "Multi-task learning as multi-objective optimization," in Advances in Neural Information Processing Systems, 2018.

[53] H. Zhang, M. Cisse, Y. N. Dauphin, et al. "mixup: Beyond empirical risk minimization," arXiv preprint arXiv: 1710.09412, 2017.

[54] S. Hou, X. Pan, C. Change Loy, et al. "Lifelong learning via progressive distillation and retrospection," in Proceedings of the European Conference on Computer Vision, 2018: 437−452.

[55] J. Deng, J. Guo, N. Xue, et al. "Arcface: Additive angular margin loss for deep face recognition," in Proceedings of IEEE Conference on Computer Vision and Pattern Recognition, 2018: 4685−4694.

[56] W. Liu, Y. Wen, Z. Yu, et al. "Large-margin softmax loss for convolutional neural networks," in Proceedings of the International Conference on Machine Learning, 2016: 507−516.

[57] F. Zenke, B. Poole, and S. Ganguli. "Continual learning through synaptic intelligence," in Proceedings of Machine Learning Research, 2017, Vol. 70: 3987−3995.

[58] C. Wah, S. Branson, P. Welinder, et al. "The caltech-ucsd birds-200-2011 dataset," Tech. Rep. CNS-TR-2011-001, California Institute of Technology, 2011.

[59] A. Chaudhry, P. K. Dokania, T. Ajanthan, et al. "Riemannian walk for incremental learning: Understanding forgetting and intransigence," in Proceedings of the European Conference on Computer Vision, 2018: 532−547.

[60] M. Riemer, I. Cases, R. Ajemian, et al. "Learning to learn without forgetting by maximizing transfer and minimizing interference," arXiv preprint arXiv: 1810.11910, 2018.

[61] L. Van der Maaten and G. Hinton. "Visualizing data using t-sne." Journal of machine learning research, 2008, Vol. 9, No. 11: 2579−2605.

[62] P. W. Koh and P. Liang. "Understanding black-box predictions via influence functions," in Proceedings of the International Conference on Machine Learning, 2017: 1885−1894.

[63] K. Deb and H. Gupta. "Searching for robust pareto-optimal solutions in multiobjective optimization," in Proceedings of the International Conference on Evolutionary Multi-Criterion Optimization, 2005: 150−164.

[64] J. -A. Désidéri. "Multiple-gradient descent algorithm (mgda) for multiobjective optimization," Comptes Rendus Mathematique, 2012, Vol. 350, No. 5−6: 313−318.

[65] T. Hospedales, A. Antoniou, P. Micaelli, et al. "Meta-learning in neural networks: A survey," arXiv preprint arXiv: 2004.05439, 2020.

[66] J. Fliege and B. F. Svaiter. "Steepest descent methods for multicriteria optimization," Mathematical Methods of Operations Research, 2000, Vol. 51: 479−494.

[67] O. Vinyals, C. Blundell, T. Lillicrap, et al. "Matching networks for one shot learning," in Advances in Neural Information Processing Systems, 2016.

[68] J. Deng, W. Dong, R. Socher, et al. "Imagenet: A largescale hierarchical image database," in Proceedings of the IEEE Conference on Computer Vision and Pattern Recognition, 2009: 248—255.

[69] P. Buzzega, M. Boschini, A. Porrello, et al. "Rethinking experience replay: a bag of tricks for continual learning," in International Conference on Pattern Recognition. IEEE, 2021: 2180—2187.

[70] A. Prabhu, P. Torr, and P. Dokania. "Gdumb: A simple approach that questions our progress in continual learning," in Proceedings of the European Conference on Computer Vision, 2020: 524—540.

[71] R. Aljundi, E. Belilovsky, T. Tuytelaars, et al. "Online continual learning with maximal interfered retrieval," in Advances in Neural Information Processing Systems, 2019.

[72] L. Risheng, L. Yaohua, Z. Shangzhi, et al. "Gradient-based editing of memory examples for online task-free continual learning," in Advances in Neural Information Processing Systems, 2021.

[73] Z. Mai, R. Li, J. Jeong, et al. "Online continual learning in image classification: An empirical survey," in Neurocomputing, 2017, Vol. 469: 28—51.

[74] Y. Li, Y. Song, and J. Luo. "Improving pairwise ranking for multi-label image classification," in Proceedings of the IEEE Conference on Computer Vision and Pattern Recognition, 2017: 3617—3625.

[75] L. Sun, S. Feng, J. Liu, et al. "Global-local label correlation for partial multi-label learning," in IEEE Transactions on Multimedia, 2022, Vol. 24: 581—593.

[76] C. Atkinson, B. McCane, L. Szymanski, et al. "Pseudo-recursal: Solving the catastrophic forgetting problem in deep neural networks," arXiv preprint arXiv: 1802.03875, 2018.

[77] J. Yoon, E. Yang, J. Lee, et al. "Lifelong learning with dynamically expandable networks," arXiv preprint arXiv: 1708.01547, 2017.

[78] J. Yoon, W. Jeong, G. Lee, et al. "Federated continual learning with weighted inter-client transfer," in Proceedings of the International

Conference on Machine Learning, 2021: 12073−12086.

[79] D. Shenaj, M. Toldo, A. Rigon, et al. "Asynchronous federated continual learning," in Proceedings of the IEEE Conference on Computer Vision and Pattern Recognition, 2023: 5055−5063.

[80] F. Lyu, W. Feng, and S. Wang. "vtgraphnet: Learning weakly-supervised scene graph for complex visual grounding," Neurocomputing, 2020, Vol. 413, No. 0: 51−60.

[81] J. Collins and J. Zimmer. "An asymmetric Arzelà-Ascoli theorem," Topology and its Applications, 2007, Vol. 154, No. 11: 2312−2322.

[82] W. A. Wilson. "On quasi-metric spaces," American Journal of Mathematics, 1931, Vol. 53, No. 3: 675−684.

[83] B. Lin, F. Ye, and Y. Zhang. "A closer look at loss weighting in multi-task learning," arXiv preprint arXiv: 2111.10603, 2021.

[84] Z. Chen, V. Badrinarayanan, C. -Y. Lee, et al. "Gradnorm: Gradient normalization for adaptive loss balancing in deep multitask networks," in Proceedings of the International Conference on Machine Learning, 2018: 794−803.

[85] X. Liu, P. He, W. Chen, et al. "Multi-task deep neural networks for natural language understanding," in Proceedings of the Annual Meeting of the Association for Computational Linguistics, 2019: 4487−4496.

[86] T. Yu, S. Kumar, A. Gupta, et al. "Gradient surgery for multi-task learning," In Advances in Neural Information Proceeding Systems, 2020.

[87] A. Chaudhry, M. Rohrbach, M. Elhoseiny, et al. "On tiny episodic memories in continual learning," arXiv preprint arXiv: 1902.10486, 2019.

[88] G. Cohen, S. Afshar, J. Tapson, et al. "EMNIST: An extension of MNIST to handwritten letters," in International Joint Conference on Neural Networks, 2017.

[89] Y. Le and X. Yang. "Tiny imagenet visual recognition challenge," CS

231N, 2015.

[90] R. Aljundi, M. Lin, B. Goujaud, et al. "Online continual learning with no task boundaries," arXiv preprint arXiv: 1903.08671, 2019.

[91] Z. Chen, J. Ngiam, Y. Huang, et al. "Just pick a sign: Optimizing deep multitask models with gradient sign dropout," in Advances in Neural Information Processing Systems, 2020.

[92] D. Silver, T. Hubert, J. Schrittwieser, et al. "A general reinforcement learning algorithm that masters chess, shogi, and go through self-play," Science, 2018, Vol. 362, No. 6419: 1140−1144.

[93] O. Russakovsky, J. Deng, H. Su, et al. "Imagenet large scale visual recognition challenge," International Journal of Computer Vision, 2015, Vol. 115, No. 3: 211−252.

[94] Z. Chen and B. Liu. "Lifelong machine learning," Synthesis Lectures on Artificial Intelligence and Machine Learning, 2018, Vol. 12, No. 3: 1−207.

[95] S. -A. Rebuffi, H. Bilen, and A. Vedaldi. "Efficient parametrization of multi-domain deep neural networks," in Proceedings of the IEEE Conference on Computer Vision and Pattern Recognition, 2018: 8119−8127.

[96] H. Li, W. Dong, and B. -G. Hu. "Incremental concept learning via online generative memory recall," IEEE Transactions on Neural Networks and Learning Systems, 2021, Vol. 32, No. 7: 3206−3216.

[97] D. -W. Zhou, H. -J. Ye, and D. -C. Zhan. "Few-shot class-incremental learning by sampling multi-phase tasks," in Proceedings of the IEEE/CVF Conference on Computer Vision and Pattern Recognition, 2022: 12816−12831.

[98] A. Douillard, A. Ramé, G. Couairon, et al. "Dytox: Transformers for continual learning with dynamic token expansion," in Proceedings of the IEEE Conference on Computer Vision and Pattern Recognition, 2022: 9285−9295.

[99] F. Zhu, H. Li, W. Ouyang, et al. "Learning spatial regularization

with image-level supervisions for multi-label image classification," in Proceedings of the IEEE Conference on Computer Vision and Pattern Recognition, 2017: 5513−5522.

[100] J. Wang, Y. Yang, J. Mao, et al. "Cnn-rnn: A unified framework for multi-label image classification," in Proceedings of the IEEE Conference on Computer Vision and Pattern Recognition, 2016: 2285−2294.

[101] F. Lyu, Q. Wu, F. Hu, et al. "Attend and imagine: Multi-label image classification with visual attention and recurrent neural networks," IEEE Transactions on Multimedia, 2019, Vol. 21, No. 8: 1971−1981.

[102] Z.-M. Chen, X.-S. Wei, P. Wang, et al. "Multi-label image recognition with graph convolutional networks," in Proceedings of the IEEE Conference on Computer Vision and Pattern Recognition, 2019: 5177−5186.

[103] T. Chen, L. Lin, X. Hui, et al. "Knowledge-guided multi-label few-shot learning for general image recognition," IEEE Transactions on Pattern Analysis and Machine Intelligence, 2020, Vol. 44, No. 3: 1371−1384.

[104] Z. Chen, X.-S. Wei, P. Wang, et al. "Learning graph convolutional networks for multi-label recognition and applications," IEEE Transactions on Pattern Analysis and Machine Intelligence, 2023, Vol. 45, No. 6: 6969−6983.

[105] K. Du, F. Lyu, F. Hu, et al. "Agcn: Augmented graph convolutional network for lifelong multi-label image recognition," in IEEE International Conference on Multimedia and Expo, 2022.

[106] C. D. Kim, J. Jeong, and G. Kim. "Imbalanced continual learning with partitioning reservoir sampling," in Proceedings of the European Conference on Computer Vision, 2020: 411−428.

[107] B. Zhou, A. Khosla, A. Lapedriza, et al. "Learning deep features for discriminative localization," in Proceedings of the IEEE

Conference on Computer Vision and Pattern Recognition, 2016: 2921−2929.

[108] D. Rolnick, A. Ahuja, J. Schwarz, et al. "Experience replay for continual learning," in Advances in Neural Information Processing Systems, 2019.

[109] Z. Mai, R. Li, H. Kim, et al. "Supervised contrastive replay: Revisiting the nearest class mean classifier in online class-incremental continual learning," in Proceedings of the IEEE Conference on Computer Vision and Pattern Recognition Workshop, 2021: 3589−3599.

[110] T.-Y. Lin, M. Maire, S. Belongie, et al. "Microsoft coco: Common objects in context," in Proceedings of the European Conference on Computer Vision, 2014: 740−755.

[111] T.-S. Chua, J. Tang, R. Hong, et al. "Nus-wide: a real-world web image database from National University of Singapore," in Proceedings of the ACM International Conference on Image and Video Retrieval, 2009: 1−9.

[112] Q.-Y. Jiang and W.-J. Li. "Deep cross-modal hashing," in Proceedings of the IEEE Conference on Computer Vision and Pattern Recognition, 2017: 3232−3240.

[113] K. Shmelkov, C. Schmid, and K. Alahari. "Incremental learning of object detectors without catastrophic forgetting," in Proceedings of the IEEE International Conference on Computer Vision, 2017: 3400−3409.

[114] G. Nguyen, T. J. Jun, T. Tran, et al. "Contcap: A comprehensive framework for continual image captioning," arXiv preprint arXiv: 1909.08745, 2019.

[115] Kingma, Diederik P. and Jimmy Ba. "Adam: A method for stochastic optimization," in International Conference on Learning Representations, 2015.

[116] M. De Lange and T. Tuytelaars. "Continual prototype evolution:

Learning online from non-stationary data streams," in Proceedings of the IEEE International Conference on Computer Vision, 2021: 8250−8259.

[117] J. Geiping, H. Bauermeister, H. Droge, et al. "Inverting gradients: how easy is it to break privacy in federated learning?" in Advances in Neural Information Processing Systems, 2020.

[118] H. B. McMahan, E. Moore, D. Ramage, et al. "Federated learning: Strategies for improving communication efficiency," arXiv preprint arXiv: 1610.05492, 2016.

[119] M. Mendieta, T. Yang, P. Wang, et al. "Local learning matters: Rethinking data heterogeneity in federated learning," in Proceedings of the IEEE Conference on Computer Vision and Pattern Recognition, 2022: 8397−8406.

[120] X. Fang and M. Ye. "Robust federated learning with noisy and heterogeneous clients," in Proceedings of the IEEE Conference on Computer Vision and Pattern Recognition, 2022: 10072−10081.

[121] B. McMahan, E. Moore, D. Ramage, et al. "Communication-efficient learning of deep networks from decentralized data," in Proceedings of the International Conference on Artificial Intelligence and Statistics, 2017: 1273−1282.

[122] F. Xiong, Z. Liu, K. Huang, et al. "State primitive learning to overcome catastrophic forgetting in robotics," Cognitive Computation, 2021, Vol. 13: 394−402.

[123] R. Aljundi, F. Babiloni, M. Elhoseiny, et al. "Memory aware synapses: Learning what (not) to forget," in Proceedings of the European conference on computer vision, 2018: 139−154.

[124] S. M. Hendryx, D. R. KC, B. Walls, et al. "Federated reconnaissance: Efficient, distributed, class-incremental learning," arXiv preprint arXiv: 2109.00150, 2021.

[125] A. Usmanova, F. Portet, P. Lalanda, et al. "A distillation-based approach integrating continual learning and federated learning for

pervasive services," in Proceedings of the IEEE Conference on Smart Computing, 2022: 86—91.

[126] D. Li and J. Wang. "Fedmd: Heterogenous federated learning via model distillation," in Advances in Neural Information Processing Systems, 2019.

[127] X. Yao and L. Sun. "Continual local training for better initialization of federated models," in IEEE International Conference on Image Processing, 2020: 1736—1740.

[128] K. Bonawitz, H. Eichner, W. Grieskamp, et al. "Towards federated learning at scale: System design," in Proceedings of Machine Learning and Systems, 2019, Vol. 1: 374—388.

[129] G. I. Winata, L. Xie, K. Radhakrishnan, et al. "Overcoming catastrophic forgetting in massively multilingual continual learning," in Annual Meeting of the Association for Computational Linguistics, 2023: 768—777.

[130] S. W. Yoon, J. Seo, and J. Moon. "Tapnet: Neural network augmented with task adaptive projection for few-shot learning," in International Conference on Machine Learning, 2019: 7115—7123.

[131] T. Lin, L. Kong, S. U. Stich, et al. "Ensemble distillation for robust model fusion in federated learning," in Advances in Neural Information Processing Systems, 2020.

[132] L. Zhang, L. Shen, L. Ding, et al. "Fine-tuning global model via data-free knowledge distillation for non-iid federated learning," in Proceedings of the IEEE Conference on Computer Vision and Pattern Recognition, 2022: 10174—10183.

[133] Y. Huang, L. Chu, Z. Zhou, et al. "Personalized cross-silo federated learning on non-iid data," in Proceedings of the AAAI Conference on Artificial Intelligence, 2021, Vol. 35, No. 9: 7865—7873.

[134] A. Santoro, S. Bartunov, M. Botvinick, et al. "Metalearning with memory-augmented neural networks," in Proceedings of the

International Conference on Machine Learning, 2016: 1842−1850.

[135] W. Huang, M. Ye, B. Du, et al. "Learn from others and be yourself in heterogeneous federated learning," in Proceedings of the IEEE Conference on Computer Vision and Pattern Recognition, 2022: 10143−10153.

[136] Y. H. Ezzeldin, S. Yan, C. He, et al. "Fairfed: Enabling group fairness in federated learning," in Proceedings of the AAAI Conference on Artificial Intelligence, 2023, Vol. 37, No. 6: 7494−7502.

[137] G. I. Parisi, R. Kemker, J. L. Part, et al. "Continual lifelong learning with neural networks: A review," Neural Networks, 2019, Vol. 113: 54−71.

[138] J. DeFauw, J. R. Ledsam, B. Romera-Paredes, et al. "Clinically applicable deep learning for diagnosis and referral in retinal disease," Nature medicine, 2018, Vol. 24, No. 9: 1342−1350.

[139] M. M. Ghassemi, T. Alhanai, M. B. Westover, et al. "Personalized medication dosing using volatile data streams," in AAAI Conference on Artificial Intelligence Workshop, 2018.

[140] C. Lee, D. Baughman, and A. Lee. "Deep learning is effective for the classification of oct images of normal versus age-related macular degeneration,"Ophthalmol Retina, 2017, Vol. 1, No. 4: 322−327.

[141] M. McCloskey and N. J. Cohen. "Catastrophic interference in connectionist networks: The sequential learning problem," in Psychology of Learning and Motivation. Elsevier, 1989, Vol. 24: 109−165.

[142] J. Luketina, M. Berglund, K. Greff, et al. "Scalable gradient-based tuning of continuous regularization hyperparameters," in Proceedings of the International Conference on Machine Learning, 2016: 2952−2960.

[143] T. Lesort, A. Gepperth, A. Stoian, et al. "Marginal replay vs conditional replay for continual learning," in International

Conference on Artificial Neural Networks, 2019: 466−480.

[144] F. Wang, J. Cheng, W. Liu, et al. "Additive margin softmax for face verification," IEEE Signal Processing Letter, 2018, Vol. 25, No. 7: 926−930.

[145] G. Hinton, O. Vinyals, and J. Dean. "Distilling the knowledge in a neural network," Computer Science, 2015, Vol. 14, No. 7: 38−39.

[146] Y. LeCun, C. Cortes and C. Burges. "The MNIST database of handwritten digits," [2024-10-10]. http://yann.lecun.com/exdb/mnist/.

[147] J. Xu and Z. Zhu. "Reinforced continual learning," in Advances in Neural Information Processing Systems, 2018.

[148] X. Tao, X. Hong, X. Chang, et al. "Few-shot class incremental learning," in Proceedings of the IEEE Conference on Computer Vision and Pattern Recognition, 2020: 12183−12192.

[149] J. Liang, R. He, Z. Sun, et al. "Distant supervised centroid shift: A simple and efficient approach to visual domain adaptation," in Proceedings of the IEEE Conference on Computer Vision and Pattern Recognition, 2019: 2975−2984.

[150] A. Ayub and A. R. Wagner. "Cognitively-inspired model for incremental learning using a few examples," in Proceedings of the IEEE Conference on Computer Vision and Pattern Recognition Workshop, 2020: 222−223.

[151] F. Lyu, S. Qing, S. Fanhua, et al. "Parallel continual learning," in Advances in Neural Information Processing Systems, 2022.

[152] K. Du, L. Li, F. Lyu, et al. "Class-incremental lifelong learning in multi-label classification," arXiv preprint arXiv: 2207.07840, 2022.

[153] R. Tiwari, K. Killamsetty, R. Iyer, et al. "Gcr: Gradient coreset based replay buffer selection for continual learning," in Proceedings of the IEEE Conference on Computer Vision and Pattern Recognition, 2022: 99−108.

[154] I. Portugal, P. Alencar, and D. Cowan. "The use of machine

learning algorithms in recommender systems," Expert Systems with Applications, 2018, Vol. 97: 205−227.

[155] M. D. Abràmoff, P. T. Lavin, M. Birch, et al. "Pivotal trial of an autonomous AI-based diagnostic system for detection of diabetic retinopathy in primary care offices," Nature Partner Journal Digital Medicine, 2018, Vol. 1, No. 1: 39.

[156] P. Chrabaszcz, I. Loshchilov, and F. Hutter. "A downsampled variant of imagenet as an alternative to the cifar datasets," arXiv preprint arXiv: 1707.08819, 2017.

[157] C. H. Lampert, H. Nickisch, and S. Harmeling. "Learning to detect unseen object classes by between-class attribute transfer," in Proceedings of the IEEE Conference on Computer Vision and Pattern Recognition, 2009: 951−958.

[158] Y. Li, S. Ren, P. Wu, et al. "Learning distilled collaboration graph for multi-agent perception," in Advances in Neural Information Processing Systems, 2021.

[159] P. Buzzega, M. Boschini, A. Porrello, et al. "Dark experience for general continual learning: a strong, simple baseline," in Advances in Neural Information Processing Systems, 2020.

[160] D. Shim, Z. Mai, J. Jeong, et al. "Online class incremental continual learning with adversarial shapley value," in Proceedings of the AAAI Conference on Artificial Intelligence, 2021, Vol. 35, No. 11: 9630−9638.

[161] B. Babcock, S. Babu, M. Datar, et al. "Models and issues in data stream systems," in Proceedings of the ACM SIGMOD-SIGACT-SIGART Symposium on Principles of Database System, 2002: 1−16.

[162] S. Teng, X. Hu, P. Deng, et al. "Motion planning for autonomous driving: The state of the art and future perspectives," IEEE Transactions on Intelligent Vehicles, 2023: 3692−3711.

[163] M. Liu, E. Yurtsever, J. Fossaert, et al. "A survey on autonomous

driving datasets: Statistics, annotation quality, and a future outlook," IEEE Transactions on Intelligent Vehicles, 2024: 1−29.

[164] M. Muhammad and G. A. Safdar. "Survey on existing authentication issues for cellular-assisted V2X communication," Vehicular Communications, 2018, Vol. 12: 50−65.

[165] M. Hasan, S. Mohan, T. Shimizu, et al. "Securing vehicle-to-everything (v2x) communication platforms," IEEE Transactions on Intelligent Vehicles, 2020, Vol. 5, No. 4: 693−713.

[166] T. -H. Wang, S. Manivasagam, M. Liang, et al. "V2VNet: Vehicle-to-vehicle communication for joint perception and prediction," in Proceedings of the European Conference on Computer Vision, 2020: 605−621.

[167] Y. Yuan and M. Sester. "Comap: A synthetic dataset for collective multiagent perception of autonomous driving," in Advances in Neural Information Processing Systems, 2021.

[168] D. Krajzewicz, J. Erdmann, M. Behrisch, et al. "Recent development and applications of SUMO—simulation of urban mobility," in International Journal on Advances in Systems and Measurements, 2012, Vol. 5, No. 3&4.

[169] Y. -C. Liu, J. Tian, N. Glaser, et al. "When2com: Multi-agent perception via communication graph grouping," in Proceedings of the IEEE conference on Computer Vision and Pattern Recognition, 2020: 4106−4115.

[170] Y. -C. Liu, J. Tian, C. -Y. Ma, et al. "Who2com: Collaborative perception via learnable handshake communication," in Proceedings of the IEEE International Conference on Robotics and Automation, 2020: 6876−6883.

[171] R. Xu, H. Xiang, X. Xia, et al. "Opv2v: An open benchmark dataset and fusion pipeline for perception with vehicle-to-vehicle communication," in Proceedings of the International Conference on Robotics and Automation, 2022: 2583−2589.

[172] S. Jain, V. K. Jain, and S. Mishra. "Vehicular traffic offloading through intelligent rsu selection in vanet," in IEEE Conference on Information and Communication Technology, 2022.

[173] D. Sunuwar and S. Kim. "Cross-layer performance evaluation of c-v2x," arXiv preprint arXiv: 2401.15844, 2024.

[174] J. Martinez, S. Pistonesi, M. C. Maciel, et al. "Multi-scale fidelity measure for image fusion quality assessment," Information Fusion, 2019: 197−211.

[175] L. Spampinato, E. Testi, C. Buratti, et al. "Madrl-based uavs trajectory design with anti-collision mechanism in vehicular networks," in Proceedings of the IEEE International Conference on Acoustics, Speech and Signal Processing, 2024: 12976−12980.

[176] Z. Xiao, Z. Mo, K. Jiang, et al. "Multimedia fusion at semantic level in vehiclecooperactive perception," in Proceedings of the IEEE International Conference on Multimedia & Expo Workshops, 2018: 1−6.

[177] Y. Maalej, S. Sorour, A. Abdel-Rahim, et al. "Vanets meet autonomous vehicles: A multimodal 3d environment learning approach," in IEEE Global Communications Conference, 2017.

[178] E. Arnold, M. Dianati, R. de Temple, et al. "Cooperative perception for 3d object detection in driving scenarios using infrastructure sensors," in IEEE Transactions on Intelligent Transportation Systems, 2020, Vol. 23, No. 3: 1852−1864.

[179] S. Thrun and J. O'Sullivan. "Discovering structure in multiple learning tasks: The tc algorithm," in Proceedings of the International Conference on Machine Learning, 1996: 489−497.

[180] H. Wang, X. Yuan, Y. Cai, et al. "V2i-carla: A novel dataset and a method for vehicle reidentification-based v2i environment," IEEE Transactions on Instrumentation and Measurement, 2022, Vol. 71: 1−9.

[181] D. Yao, M. Dai, T. Wang, et al. "Intelligent sensing and

communication assisted pedestrians recognition in vehicular networks: An effective throughput maximization approach," in IEEE Conference on Computer Communications Workshops, 2022.

[182] F. Lyu, Q. Sun, F. Shang, et al. "Measuring asymmetric gradient discrepancy in parallel continual learning," in Proceedings of the IEEE International Conference on Computer Vision, 2023: 11411−11420.

[183] Y. Cui, L. Yang, and H. Yu. "Learning dynamic query combinations for transformer-based object detection and segmentation," in Proceedings of the International Conference on Machine Learning, 2023: 6591−6602.

[184] H. Xie, J. Zhu, M. Khayatkhoei, et al. "A critical view of vision-based long-term dynamics prediction under environment misalignment," in International Conference on Machine Learning, 2023: 38258−38271.

[185] E. Verwimp, K. Yang, S. Parisot, et al. "CLAD: A realistic continual learning benchmark for autonomous driving," in ACM Multimedia, 2022: 659−669.

[186] K. Huang, F. Wang, Y. Xi, et al. "Prototypical kernel learning and open-set foreground perception for generalized few-shot semantic segmentation," in Proceedings of the IEEE International Conference on Computer Vision, 2023: 19256−19265.

[187] Y. Bi, Q. Liu, S. Zhu, et al. "Design and analysis of dual three-phase winding pmsm for integrated ema," in IEEE International Electrical and Energy Conference, 2021.

[188] M. Kang, J. Zhang, J. Zhang, et al. "Alleviating catastrophic forgetting of incremental object detection via within-class and between-class knowledge distillation," in Proceedings of the IEEE International Conference on Computer Vision, 2023: 18894−18904.

[189] Y. Chen. "Cross-scale dilated residual network for image compressed sensing," in Proceedings of the International Conference on

Communications, Information System and Computer Engineering, 2023: 174−178.

[190] E. Xhoxhi, V. A. Wolff, Y. Li, et al. "Vulnerable road user clustering for collective perception messages: Efficient representation through geometric shapes," arXiv preprint arXiv: 2404.14925, 2024.

[191] J. Huo and T. L. van Zyl. "Comparative analysis of catastrophic forgetting in metric learning," in Proceedings of the International Conference on Soft Computing & Machine Intelligence, 2020: 68−72.

[192] Y. Jiang, E. Javanmard, J. Nakazato, et al. "Roadside lidar assisted cooperative localization for connected autonomous vehicles," in IEEE/RSJ International Conference on Intelligent Robots and Systems, 2023.

Acknowledgement

First and foremost, we would like to express my deepest gratitude to the National Science and Technology Academic Publication Foundation of China for their major financial support, which has made the successful implementation of this project possible.

This book was also supported by National Natural Science Foundation of China (Grant No.62476189, 62406323, 62373355), Suzhou Science and Technology Development Plan Project (Grant No.SS202133), Postdoctoral Fellowship Program of CPSF under Grant Number GZC20232993, China Postdoctoral Science Foundation (Grant No.2024M753496) and Jiangsu Province Graduate Research and Practical Innovation Plan (Grant No. KYCX23_3319). We are particularly grateful to our students, Kaile Du and Chenggong Ni, for their help in compiling and formatting this book. Their meticulous attention to details has ensured that the format is correct and the book is more accessible to readers. We would also like to acknowledge the publisher, Soochow University Press, for their professional assistance and for believing in the value of this work. We would like to express our sincere thanks to the Suzhou Key Laboratory of Intelligent Low Carbon Technology Application, Jiangsu Industrial Intelligent Low Carbon Technology Engineering Center, School of Electronics and Information at Suzhou University of Science and Technology, the New Laboratory of Pattern Recognition at the Institute Automation, Chinese Academy of Sciences, Suzhou College of Economics and Trade, and Tianjin University for providing the research facilities that enabled our research to proceed smoothly.

About the Authors

Linyan Li is an Associate Professor with the Suzhou Institute of Trade & Commerce, Suzhou, China. She received the master's degree from Wuhan University, Wuhan, China, in 2007. Her current research interests include machine learning, neural information processing, and pattern recognition.

Fuyuan Hu is a Professor with School of Electronic & Information Engineering, Suzhou University of Science and Technology, Suzhou, China. He was a Postdoctoral Researcher with Vrije Universiteit Brussel, Brussels, Belgium, a PhD student with Northwestern Polytechnical University, Xi'an, China, and a visiting PhD student with the City University of Hong Kong, Hong Kong. His research interests include machine learning, continual learning and computer vision.

Fan Lyu is a Postdoc with New Laboratory of Pattern Recognition, Institute of Automation, Chinese Academy of Sciences. He received the BS and MS degree in School of Electronic & Information Engineering, Suzhou University of Science and Technology, China, in 2015 and 2018. He got his PhD degree in the College of Intelligence and Computing, Tianjin University, in 2023. His research interests include sustainable AI and long-term AI.